EVIDENCE
MERCURY CAUSES
ALZHEIMER'S DISEASE

"THE DENTAL AMALGAM STORY"

EVIDENCE MERCURY CAUSES ALZHEIMER'S DISEASE

"THE DENTAL AMALGAM STORY"

ROBERT SIBLERUD

THE REGENCY PUBLISHERS

Library of Congress Control Number: 2021922148

ISBN: 978-1-957054-00-1   (Paperback Edition)
ISBN: 978-1-957054-01-8   (Hardcover Edition)
ISBN: 978-1-956736-99-1   (E-book Edition)

Book Ordering Information

The Regency Publishers, US
521 5th Ave 17th floor NY, NY10175

Phone Number: (315)537-3088 ext 1007
Email: info@theregencypublishers.com
www.theregencypublishers.com

Printed in the United States of America

TABLE OF CONTENTS

FOREWORD

For over a century, scientists have been searching for the cause and treatment of Alzheimer's Disease (AD). Could it be as simple as mercury toxicity? After you finish reading this book, I think you will agree that is probably the cause. Where is the mercury coming from? Governmental organizations tell us it is from fish, air (contaminated by coal burning power plants), and some medical treatments including vaccines and dental amalgams. According to the World Health Organization, the greatest source of mercury exposure originates from silver dental fillings called amalgams, which contain 50% mercury. Mercury vapors are continuously being released from amalgam fillings and inhaled. It enters the blood stream from the lungs, and then crosses the blood-brain barrier and into brain cells where it becomes ionized and entrapped for up to several decades.

Dr. Siblerud's earlier research showed that the mental health of young adults with amalgams is significantly worse compared to a control group without amalgams. This suggests that mercury is already causing changes in the brain at a young age. As you will see in this book, there is anecdotal evidence of some people recovering from AD following amalgam removal and detoxification. Of course not everyone does recover. I had one AD patient who had his dental amalgams removed as well as root-filled teeth plus other metals. He was also detoxified but showed no improvement. On the other hand, I had some amyotrophic lateral sclerosis patients (ALS, Lou Gehrig's disease) that improved following amalgam removal. The disease progress stopped and the patients got better following removal of amalgams and other dental metals. The ALS patients were detoxified and shielded from electromagnetic fields, and given healthy foods and supplements. There are some similarities between Parkinson's Disease and AD, with many Parkinson's Disease patients developing AD. A number of my patients with Parkinson's Disease got better after the treatment. To detoxify mercury, they were given DMPS (ca. 100), DMSA, and later Penicillamin or Tiopronin and Irminix (Chapter Eight).

As you read in this book, there are many health disorders that improve or are eliminated following amalgam removal, and you will begin to understand that mercury is one of the most toxic elements there is and can cause a multitude of health disorders where the cause is not known.

Then you will ask yourself, how was this allowed to happen? This is the saddest part of the story. You will discover in the book about three "amalgam wars" beginning in the 1800s, where enlightened dentists have tried to stop the dental profession from using mercury because they knew how toxic mercury was.

Around 1980, the "third amalgam war" began, which was started by a Colorado dentist named Dr. Hal Huggins (Chapter Nine) who found many health disorders that improved or were eliminated following amalgam removal. He reports a number of his Alzheimer's disease patients recovered following amalgam removal. Other successes include multiple sclerosis patients, Parkinson's disease patients, and ALS patients to name a few other disorders.

Organized dentistry rewarded him by revoking his dental license. Today, in most states in the U.S., if dentists mention that dental amalgams are harmful, they can have their license revoked. Why is this cover-up happening? One of the dental professions greatest fears is liability for all the health problems they have caused by using mercury fillings. This sounds similar to the tobacco industry. Governments will not fund research regarding amalgam mercury toxicity. Dr. Siblerud's research was funded by a private source. Most scientific journals are reluctant to publish articles regarding amalgam mercury toxicity, no matter how good the research was, again trying to preserve status quo. In the meantime, many people have suffered from illnesses that could have been prevented resulting in countless dying indirectly from amalgam mercury toxicity.

My own life was almost destroyed when I was 18 years old working as an educated electrician. I suffered from severe amalgam mercury toxicity. I had diarrhea, tiredness, insomnia, burning feet, tinnitus, tremor, too much saliva, severe depression, cold feet and hands, weakness, hair loss, and many more symptoms. I began to recover after amalgam removal and a change in nutrition along with detoxification of heavy metals, I lost 30 kg of weight. All of my back teeth had been filled with mercury amalgams since childhood. I then studied medicine and became an environmental medicine physician.

I worked as a physician for over seven years at the University Medical Center of Freiburg (Germany) becoming interested on how mercury may cause AD and other illnesses. I had a number of scientific papers published and was involved in research that could explain how certain genes in AD that were involved with apolipoproteins, which could increase the risk of AD. Our research found that the apolipoprotein E2 allele could prevent AD, while the apolipoprotein E4 allele could increase the risk of AD. The E2 allele has two cysteines that can remove mercury from the cell. The E4 allele could not remove mercury from the cell because it had no cysteines. I also co-authored a paper with Dr. Siblerud explaining how the physiological and pathological aspects of AD could be explained by mercury toxicity (chapter 9).

Many dentists have stopped using dental amalgams, and the AD rate is declining. Several European countries have also banned dentists from using mercury amalgams.

This book provides so much evidence that shows how mercury from dental amalgams is causing AD and many other health disorders. Hopefully, it will provide the stimulus to enact legislation around the world to ban the use of dental mercury amalgams.

Dr. Joachim Mutter
Environmental Physician

PREFACE

This book is about my journey into the scientific world trying to find the cause of unexplained illnesses. Originally, I was trying to find the cause of mental health problems. I had a loved one who developed a mental illness and was helped by nutrition and supplements recommended in alternative health literature. I then asked myself, why weren't all psychotherapists using nutrition in their practices? I was upset about this and decided the only way to change the system was to get credentials. I entered the masters' program in nutrition at Colorado State University. Because I was still practicing optometry, I did my thesis research on nutrition and myopia, and found a very significant hair mercury finding that I could not explain.

About that time, a dentist named Dr. Hal Huggins from Colorado Springs was claiming that mercury from silver dental fillings (amalgams) was causing all kinds of health problems. I wondered if my hair mercury finding had anything to do with amalgam mercury. Because my office was located across from campus, and it was easy to attend classes and get back to my practice, I entered the Ph.D. program in physiology. I decided to do my dissertation on health effects of dental amalgams. Dr. David Robertshaw, the head of the physiology department, was my major professor and agreed to mentor the research.

My research found many health problems that could be associated with amalgam mercury, including a number of mental health problems. Nutrition was a means of neutralizing mercury toxicity, as I later found out.

To my dismay, I found a world of politics in the scientific community trying to preserve status quo, and then a world of denial by the dental community ignoring the science of amalgam mercury toxicity. The dental boards threatened to revoke licenses of dentists who spoke of the harmful effects of dental amalgams. I am sure this denial was based on fear of litigation, if it could be proven that amalgam mercury was harmful.

In the meantime, millions of people in America and worldwide have suffered countless health problems, including neurological disease, mental health problems, and Alzheimer's disease, which has claimed millions of lives over the past four decades since this problem of amalgam mercury toxicity was brought to the public's awareness.

The dental schools have known about the hazards of amalgam mercury and refuse to do research, including the University of Colorado dental school who helped fund my Ph.D. research in the 1980s. Colorado State University illegally dismissed me from the Ph.D. program after I completed all the required course work and agreed-upon research. They were trying to prevent me from publishing my research findings regarding amalgam mercury toxicity. (I brought a lawsuit against CSU and the judge ruled that I was illegally dismissed.) In the meantime, countless people have suffered from dental mercury poisoning. Since then, I have had 20 scientific papers published about amalgam mercury toxicity.

I received a grant to study Alzheimer's disease and its relationship to amalgam mercury in the 1990s. The Alzheimer's Association refused to cooperate in recruiting patients for the study stating that mercury did not cause Alzheimer's disease, again trying to preserve the status quo. As a result, I was unable to complete the study regarding mercury and Alzheimer's disease. It was when I was researching the scientific literature for a book I was writing about preventing Alzheimer's disease that I could explain 70 pathological and physiological changes occurring in Alzheimer's disease that could be caused by mercury toxicity.

Then I knew, I had the scientific proof. To get the word out, I needed to write a scientific paper about my findings and a book explaining how dental amalgam mercury was causing Alzheimer's disease. The paper was submitted to a number of top scientific journals but was turned down. Finally, it was accepted by a top rated peer-reviewed journal, the International Journal of Environmental Research and Public Health. I have discovered that the scientific community does not want to get involved in alternative health care even though there is good science backing it up. Otherwise it would disrupt the status quo and possible funding for their own research.

Hopefully, this book will stimulate further research regarding Alzheimer's disease and its relationship to mercury toxicity. There is enough scientific evidence to ban dental amalgam mercury in America. A number of European countries have banned amalgams, and Alzheimer's disease has decreased in these countries. Many American dentists have also curtailed the use of amalgams, and I would like to thank them on behalf of humanity. The overall use of amalgams in the United States by dentists has declined as has the Alzheimer's rate.

I still wonder why the dental profession was allowed to place one of the most toxic elements in people's mouths for nearly two centuries. I personally believe this has been a crime against humanity. Form your own opinion after reading the book, and I am sure you will agree with me.

Dr. Robert Siblerud
Environmental Physiologist

CHAPTER ONE

Alzheimer's Disease 101

Understanding the Malady

As you will discover in this book, the cause of Alzheimer's disease (AD) may literally be right under our nose, the mercury in dental fillings. Before we explain how mercury may be causing AD, it is important to understand the disease itself. Alzheimer's disease was first identified in 1906 by Alois Alzheimer, a German psychiatrist, who performed an autopsy on a demented female patient. A year later, he reported his findings of abnormal waxy fragments (amyloid plaques) and twisted fibers (tau tangles) that now define the disease. Since his discovery over a century ago, the number of AD patients being diagnosed each year has increased, where AD is now the fourth leading cause of death in the United States, killing over 120,000 people each year.

Alzheimer's disease is the most common of 70 types of dementia, and some health care experts predict that by 2050, approximately 2 billion people worldwide will suffer from dementia, costing about $1 trillion in medical expenses annually. In 2010, worldwide costs of the disease totaled $604 billion. In the United States, approximately 5 million people suffer from AD, at an annual estimated cost of $250 billion. The number of Americans expected to be diagnosed with AD is projected to reach 14 million by 2050. World wide, the Alzheimer's Disease International estimates the number of people affected by dementia will exceed 80 million people. By 2050, it is estimated that 22 percent of the world's population will be older than 60 years. About one in ten individuals over the age of 65 and one half over the age of 85 will develop dementia.

The statistics are frightening. However, one recent study reported that the rate of AD may be on the decline. This is encouraging and may be related to a decrease in mercury exposure as we will discuss later. There are a number of methods that can reduce the risk of AD and delay its onset, and these will be discussed in the last chapter. But first, let's discuss the most common types of dementia.

TYPES OF DEMENTIA

Alzheimer's disease is the most common of the seventy plus types of dementia. Only after autopsy, showing the presence of plaque and tangles, can AD be definitely diagnosed. It is important to identify the dementia types to insure the proper treatment for the memory loss. Mercury has also been suspected in other types of dementia.

Lewy Body Disease (LBD): LBD accounts for 10 to 20 percent of mild cognitive impairment and dementia in older adults. It is a degenerative disorder associated with abnormal tissue called Lewy bodies found in certain areas of the brain. Lewy bodies contain certain deposits of protein called alpha-syncline, that are associated with Parkinson's disease. The neurotransmitters acetylcholine and dopamine are reduced in the brain of Parkinson's disease patients. Early symptoms include confusion, daytime drowsiness, sleep disorders, attention difficulties, poor concentration, and short term memory difficulties. Physical movements similar to Parkinson's disease may occur and possible visual hallucinations. In later stages, an MRI can detect brain atrophy.

Parkinson's Disease (PD): PD is caused by loss of neurons that make dopamine in parts of the brain that control muscle activity such as walking, standing, speaking, and eating. If a PD patient lives to be 85 years, there is a 67 percent chance of dementia. The chance of becoming demented increases 5 percent for every year the person has PD. Dementia comes three or more years after movement-related symptoms appear. Evidence suggests there is also a relationship between mercury and PD that will be discussed later.

Frontal Temporal Lobe Dementia (FTLD): FTLD affects the frontal and temporal lobes of the brain. It is associated with deep areas of the brain being affected, including basal ganglia and the hypothalamus that regulates hormones. FTLD accounts for 10 percent of all dementia and often begins in people under 65 years. As the disease progresses, it causes an irreversible decline within 2 to 15 years. The most common cause is abnormality of the tau protein that provides structure to the neuron. The tau protein twists, causing the neuron to die in a tangled mass. Mercury can cause this.

Early symptoms may include apathy, loss of normal social inhibitions such as saying anything one thinks, inability to understand or use language, reduction in the amount of talking, inability to perform complex tasks, and senseless actions such as waving goodbye. During the early stages, there is a normal MRI test and blood profile. A SPECT scan may show mild reduced activity. ***Vascular Dementia (VD)***: VD results from an insufficient blood flow to brain areas causing cognitive problems. This reduced flow can be caused by hypertension, diabetes, heart disease, stroke, smoking, excess alcohol, and a sedentary lifestyle. This form of dementia is very common and accounts for 15 to 25 percent of all cases of dementia. It can develop suddenly or gradually. If blood vessels fail to deliver blood containing oxygen and glucose, the cells can be damaged within minutes. Accumulation of unrecognized small strokes is the most common cause.

Early symptoms may include impaired attention, apathy, impaired judgment, taking longer to perform tasks, difficulty in expressing words, walking problems, and clumsiness.

Head Injury Dementia: Nearly two million head injuries occur each year. For those carrying the APOE 4 gene, such an event is very dangerous. If people have this gene, the risk of developing AD increases two fold. For those who do not have this gene, the risk does not increase.

Infections and Immunological Causes of Dementia: Infections by prions, viruses, bacteria, and fungi may invade the brain and not be contained by the immune system. The most common infection may occur in multiple sclerosis patients where the myelin sheath of the nerves has been damaged. It may also occur in chronic fatigue syndrome caused by viruses. The herpes virus, cytomeglo virus, HIV virus, and bacterial infections of the heart may cause dementia.

Alcohol Related Dementia and Other Brain Toxins: The most common toxins causing dementia may be addictive drugs such as sedatives, cocaine, and speed. Alcohol, anti-convulsive drugs, toxic fumes, industrial chemicals, and heavy metals such as mercury, may also cause dementia.

Large amounts of alcohol, four or more glasses of wine or equivalent hard liquor daily may increase the risk of these dementias. However, a glass of wine, one time a week, may significantly reduce these dementias by up to 70 percent.

Mercury may be associated with a variety of these dementias but Alzheimer's disease is the most researched.

THE SEVEN STAGES OF ALZHEIMER'S DISEASE

Stage I: During the earliest stages of AD, there is no cognition or memory impairment. Changes in the brain can begin 20 to 30 years before the first symptom.

Stage II: Very mild cognitive decline occurs in Stage II AD. Memory lapses occur, such as forgetting familiar words, names, or locations of everyday objects.

Stage III: Mild cognitive impairment occurs during this stage, which is noticeable to family, friends or co-workers. Symptoms may include problems with familiar names, a decreased ability to remember familiar information, inability to retain new information, performance issues at work, and losing or misplacing valuable objects.

Stage IV: Moderate cognition decline with clear deficiencies such as decreased knowledge of recent occasions, impaired ability to perform challenging mental exercises such as counting backwards from 100 by 7s, decreased ability to pay bills or shopping for groceries and managing finances, and appearing subdued and withdrawn in social situations. A diagnosis of AD can be made with certainty from the beginning of this stage.

Stage V: This stage, moderate AD, finds moderately severe cognition decline with major gaps in memory and deficits in cognition function. AD patients can't recall important details such as addresses and phone numbers. There is confusion about days, dates, and seasons. Counting backwards from 20 by 2s is difficult. Subjects have difficulty in doing abstract reasoning such as dressing appropriately for the season.

Stage VI: This is the moderately severe stage of AD, which lasts about 2.5 years. Memory difficulties worsen with significant personality changes. There is a need for assistance with routine daily activities such as bathing, dressing, and toiletry. Many will be incontinent. Often there is a fear of being alone. They will have difficulty counting backwards from 10 by 1s. Verbal outbursts will occur in this stage.

Stage VII: The final stage finds severe cognitive decline. The AD patient ultimately can't control movement, and speech is affected. Speech is limited to about six intelligible words. The ability to ambulate is lost with many falling over even when seated. Physical rigidity sets in, and 40 percent manifest deformities. During the final stage, death occurs often by pneumonia caused by aspiration. In stage VII, the AD patient is vulnerable to stroke, cancer, and heart disease.

RISKS FOR ALZHEIMER'S DISEASE AND DEMENTIA

Some people are more at risk of developing AD than others, as indicated by research. By knowing these risks one can possibly delay the onset of AD. Often this means a change in lifestyle.

Age: One risk you can't control is age, which is the greatest risk for developing AD. The odds of developing AD under 65 is 1 in 20. Over the age of 65, the risk increases to 1 in 14, and 1 in 6 over the age of 80.

Gender: For most types of dementia, men and women are at equal risk, However, women are more likely to develop AD than men. In Great Britain, AD is the biggest killer of women and third largest killer for men. For mild cognitive impairment, men are at higher risk than women.

Estrogen: After menopause, lower levels of estrogen can lead to diminished cognition. Women are more likely to carry the APOE

4 gene that increases their AD risk.

Ethnicity: African Americans and Hispanics have a higher overall risk of developing AD.

Epigenetics: Epigenetics is the scientific study that examines how the environment affects genes. There are many factors that can influence whether a gene is expressed, whether it is switched on or off. Diet and lifestyle are two of the biggest factors. Also exposure to toxins and smoking can activate a gene.

Genes: The gene that has the strongest influence on the risk of AD is the apolipoprotein (APOE) gene located on chromosome 19. APOE carries cholesterol and supports the transport and clearance of beta amyloid protein found in AD. There are three types of APOE genes called alleles, consisting of E2, E3, and E4. A person inherits one of these genes from each parent. There are six possible combinations. (1) E2 and E2, (2) E2 and E3, (3) E2 and E4, (4) E3 and E3, (5) E3 and E4, and (6) E4 and E4. APOE 2 is mildly protective against AD. However, if you have one E2 and one E4, the risk increases 2.6 times for developing AD. An E3 and E4 combination increases the risk by 3.3 fold, and if two E4, the risk increases by a factor of 14.9 times.

Early onset AD occurs between the ages of 30 and 60 and is of genetic origin. According to the Alzheimer's Association, in the United States, genetics are responsible for less than 5 percent of AD. About 20 genes have been identified as genes that increase the risk of AD, even though they do not cause it.

Head Injuries: Head injuries can cause chronic traumatic encephalopathy. A blow to the head increases the risk of AD because brain cells are damaged, especially if one has an APOE4 gene.

The head circumference and brain size are inversely associated with AD.

Diabetes: In type 2 diabetes, the message to the pancreas is not working properly. It continues to produce insulin but insulin receptors become resistant resulting in high blood sugar, in which case, the risk of developing AD increases 50 to 65 percent.

Sugar: Studies have shown that high levels of blood glucose can increase levels of beta amyloid protein found in AD. Mouse studies found doubling blood glucose will increase brain beta amyloid protein by 20 percent.

Cardiovascular Disease: Heart disease is believed to double the risk for AD and vascular dementia. Brain cells use at least 20 percent of the food and oxygen that the blood carries. Research suggests that plaque and tangles are more likely to cause AD if strokes damage the brain blood vessels.

Hearing Loss: A study at John Hopkins School of Medicine studied 639 subjects aged 36 to 50 for a period of 18 years. Initially, none had memory impairment. Those who developed mild, moderate, and severe hearing loss were two times, three times, and five times more likely to develop dementia. The theory is the brain becomes so fatigued that it loses its effective processing power. Those who have hearing loss become socially isolated. The worse the hearing loss, the greater the risk.

Gum Disease: Periodontal gum disease is associated with a six-fold increase in the rate of cognitive decline in early stage AD. Gum disease produces high levels of inflammation that may affect the brain. Chronic inflammation has a detrimental effect on the progression of AD.

Proton Pump Inhibitor (PPI): PPI is a medication that reduces acid reflux. This medication has been found to increase the risk of dementia by 44 percent. PPI increases the level of beta amyloid protein found in AD. It is thought that PPI reduces the ability of the body to absorb zinc.

Over the Counter Medicine (OTC): Research has shown OTC anti-cholinergic medicine, which blocks the neurotransmitter acetylcholine, reduces the brain volume and creates cavities in the brain resulting in memory loss.

Other Risk Factors: Depression, anxiety, sleep apnea, HIV infection, and chronic kidney disease all increase the risk for dementia.

Lifestyle: People with the healthiest habits in mid life have a lower risk of dementia later on in life, especially if they do three or more of the following: 1. Exercise regularly 2. Avoid smoking 3. Moderate alcohol consumption, and 4. Maintaining a healthy weight with a good diet.

Diet: Too much salt, sugar, and saturated fat increases dementia risk. ***Alcohol:*** It was thought low to moderate alcohol consumption might reduce the risk of dementia. Some research now suggests that one drink a week or month might decrease the risk.

Fitness: Physical inactivity is one of strongest lifestyle risk factors. One study found memory improved after only four weeks of exercise (One hour three days a week to include aerobic, strength, and stretch exercises).

Stress: Stress has a negative impact on memory and concentration.

Sleep: During sleep, beta amyloid protein is cleared from the brain where plaque is built up.

Smoking: Smoking increases the risk of dementia by 45 percent. One study suggested that 14 percent of AD may be the result of smoking.

Inflammation: Inflammation is involved with AD and is one of the first responses of the immune system.

Education: Lower intelligence and limited education increases the risk of AD.

Mediterranean Diet: One study followed 2258 Medicare recipients over 14 years. One third of them who followed the Mediterranean diet were 39 percent less likely to develop AD.

A SYNOPSIS OF SOME RISK FACTORS FOR AD AND THEIR POTENTIAL TO BE ASSOCIATED WITH AD

- One family member with AD increases the risk by 3.5 times.

- More than one family with AD increases the risk 7.5 times.

- A family history of Down's Syndrome increases the risk by 2.7 times.

- A single-head injury with loss of consciousness doubles the risk.

- Alcohol or drug dependence increases the risk 4.4 times.

- Major depression doubles the risk.

- Stroke increases the risk tenfold.

- High cholesterol increases the risk 2.1 times.

- Heart disease increases the risk by 2.5 times.

- High blood pressure increases the risk by 2.3 times.

- Diabetes increases the risk 3.4 fold.

- History of cancer and cancer treatment increases the risk three fold.

- A history of seizures increases the risk by 1.5 times.

- Exercising less than two times a week doubles the risk.

- Those with less than a high school education double their risk.

- Jobs not requiring new learning double the risk.

- Those aged between 65 and 74 double the risk.

- Those aged 75 to 84 increase the risk by seven times.

- Those over 85 have a risk of 38 times more likely to develop AD.

- Smoking cigarettes for ten years or longer increases the risk by 2.3 times.

- Those with one APOE 4 gene increases the risk by 2.5 times.

- Those with two APOE 4 genes increase the risk fivefold.

THE BRAIN OF ALZHEIMER'S DISEASE

Alzheimer's disease (AD) can begin up to 30 years before any symptoms manifest. All this time, there are changes going on within the brain. The first changes occur in the limbic region of the brain, which includes the hippocampus. The hippocampus is the control switchboard for the memory system. It is responsible for sorting pieces of information, storing them in different parts of the brain, and recalling them when you need the information. In addition, AD attacks the amygdala in the limbic system, which governs emotions such as anger, fear, and the fight and flight response. The disease then spreads to the frontal and temporal lobes of the cerebral cortex.

AD then destroys neurons deep within the brain called the basal nucleus of Meynert, which is rich in the neurotransmitter acetylcholine that forms and retrieves memory.

Like mercury toxicity, AD also decreases the following neurotransmitters including dopamine, involved with movement; glutamate, involved in learning and remote memory; norepinephrine, involved in emotional response; and serotonin involved with mood and anxiety. Mercury also suppresses acetylcholine.

In AD, a substance called beta amyloid protein begins to aggregate, which then precipitates to form plaque-like deposits at specific sites of the brain. These cloud-like deposits are formed in deep cortical layers. The beta amyloid protein is snipped off by enzymes from the amyloid precursor protein, which play a role in neural growth. Various kinds of protein fragments are snipped off by the amyloid precursor protein that appear to be longer and stickier than others. These protein fragments collect together and harden to form a plaque unlike other protein fragments that dissolve easily and are removed from the brain. The buildup of beta amyloid occurs early in AD and may develop many years before signs of cognitive decline. As will be discussed later, mercury can produce plaques similar to AD plaques.

The other hallmark pathology found in AD is the progressive deposit of abnormal phosphorylated tau protein in susceptible nerve cells that lead to their destruction. A very noticeable change occurs in the distal segment of the nerve dendrites, which follow a tortuous course. Slender threads appear in the altered distal dendrites resulting in the formation of neurofibrillary tangles (NFT).

Tangles caused by the breakdown of tau protein occur inside a cell body. Tau protein undergoes changes in AD that cause the cell to malfunction. The strand of tau unravels and clumps together to form tangled masses inside the cell. The internal cell structure made up of microtubles collapse and disrupts nutrient transport and electrical impulse transmission leading to breakdown of cell function.

Tangles and plaques appear firmly intertwined in the disease process, but scientists cannot explain the mystery that causes them. As will be discussed in another chapter, animal studies have shown that mercury can cause both tangles and plaques similar to AD.

Oxidative Stress

Following damage to certain cellular structures, oxidative stress occurs within the cell body of mitochondria, which provides energy to the cell. Damaged mitochondria tend to produce highly reactive molecules called free radicals that can cause tissue breakdown and DNA damage.

What causes free radicals? Normal aging can cause them, as well as inflammation, and mercury toxicity. Signs of oxidative stress have been observed in AD brains in later stages of plaques and tangles. Some researchers believe that it is the oxidative stress that leads to formation of plaques and tangles. Low levels of oxidative stress may be sufficient to damage the brain neuron.

Inflammation

Inflammation is the body's natural response to injury, resulting in pain, swelling, heat, and redness. Studies have shown that inflammation develops in the AD brain. Microglia are immune cells that clear out dead cells and waste products from the brain. Scientists speculate that microglia identify plaques as foreign substances and try to destroy them, resulting in inflammation. Microglia may be damaging neurons by activating immune compounds such as interleukin-1 and an enzyme called Cox-2 that causes inflammation. Mercury is known to cause inflammation.

Vascular Brain Injury

Some researchers believe that a faulty vascular system in the brain may create an environment for neuron destruction. Almost all people with severe cerebrovascular disease have an abundance of amyloid beta protein. Many people with AD also have cerebrovascular disease.

Insulin Resistance

Scientists cannot explain why Alzheimer's disease and diabetes are related. People with type 2 diabetes are at a higher risk for developing AD. Type 2 diabetes affects the ability of the brain and other tissue in the body to use sugar and respond to insulin. Too much glucose accumulates in the blood. It results in a decrease of energy metabolism in the brain, while increasing inflammation and oxidative stress. An imbalance of insulin may affect the clearance of amyloid beta protein and increase the amount of tau protein. Chronic high blood glucose can trigger the production of advanced glycation, a toxic molecule in the brain. Mercury has an affinity for sulfhydryl groups that are found in insulin, perhaps explaining the insulin imbalance.

Genetic

Familial forms of Alzheimer's Disease account for 15 percent of all AD worldwide. The sporadic form, late onset AD, accounts for 85 percent. A family history of AD increases the risk by four fold. The apolipoprotein E4 allele is a marker for both familial and sporadic AD. Mutations of the amyloid precursor protein gene and the two presenilin genes account for five percent of all AD.

There are four mutated chromosomes that can cause AD.

1. AD is induced by a mutation of the beta amyloid precursor protein gene on chromosome 21.

2. There are 43 mutations of the presenilin 1 gene on chromosome 21 in early onset AD.
3. Two mutations of the presenilin 2 gene occur on chromosome 1 in early onset AD.
4. In late onset AD, the genetic risk is related with the gene on chromosome 19 encoding apolipoprotein E.

Apolipoproteins are macromolecule complexes that carry lipids such as cholesterol, triglycerides, and phospholipids from one cell to another within a tissue or between organs. Apolipoprotein E (APOE) is synthesized by microglial cells, predominantly astrocytes. APOE plays a role in remodeling and sprouting neurons in response to tissue injury. It maintains nerve integrity and cholinergic activity. APOE helps in the development, maturation, and repair of cell membranes of neurons. Three alleles, E2, E3, and E4 are found at a single gene locus on chromosome 19. APOE 4 has been found to be over-represented in both familial and late onset AD. APOE 4 was found to be three-fold higher in 40 to 50 percent of AD patients. The E2 allele is considered to be protective. For late age onset AD, the mean age of onset decreases from 78 to 70 if one has the E4 allele. The number of APOE 4 alleles correlates with senile plaque and neurofibrillary tangles (NFT) in late onset AD. People inherit one type of the three APOE alleles from each parent with six possibilities of allele combinations. People without APOE 4 have an estimated risk between 9 and 20 percent of developing AD by the age of 85. Those with two copies increase the risk to between 50 and 90 percent, while one copy increases the risk to between 25 and 60 percent.

The amyloid precursor protein (APP) undergoes a series of proteolytic cleavages resulting in the production of small amino acids referred to as beta amyloid peptides. APP maintains cholesterol homeostasis during nerve reinnervation and synaptic replacement. A mutation to the APP gene causes an over production of the neurotoxic form of beta amyloid, which leads to plaques. Researchers are undecided whether amyloid plaque deposits are responsible for the events leading to cell death in AD.

Mutations to presenilin 1 gene are associated with the age of onset in AD, usually manifesting in the late 30s or early 40s. The mutation leads to the overproduction of beta amyloid protein in the brain and fibroblasts. Mutations of the presenilin 2 gene affect the role of a receptor in trafficking and partitioning proteins inside the cell and is associated with early onset AD.

Alzheimer's Disease has been a great mystery to scientists for over a century. What causes these neuritic plaques and neuro-fibrillary tangles? Evidence has been presented that there are many risk factors for developing AD, which can be reduced by certain lifestyle changes. Even if one has a genetic predisposition for AD, lifestyle changes can greatly reduce the risk. This evidence suggests there may be an environmental etiology factor in AD. Evidence will be presented that all these neurological/pathological changes in AD can be explained by mercury toxicity. Before we get into the specifics on how mercury may be causing AD, we need to understand why mercury is one of the most toxic elements on the planet.

CHAPTER TWO

Mercury

A Very Toxic Element

Before we discuss how mercury may be causing Alzheimer's disease, it is important that we understand why this element is one of the most toxic elements on Earth. The World Health Organization (WHO) describes mercury as the third most toxic element stating, "There is no safe level of mercury." Yet the dental profession is still using dental amalgam fillings that contain 50 percent mercury, and which is continuously being released as vapor that enters the blood stream after being inhaled into the lungs. Most people in industrial countries have amalgam dental fillings. The healthcare profession is still using mercury in medications as a preservative.

In 1991, the WHO stated the largest average intake of mercury by humans originates from the dental amalgam filling. Because mercury is the most toxic of non-radioactive elements, it can cause a multitude of health problems including neurodegenerative disease. Mercury is ten times more toxic when acting on neurons than lead, which is another neurotoxin. Erethism and Minamata disease are two conditions caused by mercury poisoning.

MERCURY DISEASES

Erethism (Mad Hatter Disease)

Erethism is a neurological disorder that affects the whole central nervous system. The disorder is characterized by behavioral changes such as irritability, low self-confidence, depression, apathy, shyness, and timidity. With prolonged exposure to mercury vapor, there is memory loss, delirium, temper, and personality changes, with difficulty in social interactions. Physical symptoms may include a decrease in physical strength, headaches, general pain, and tremors.

Erethism was common among old England felt-hat makers who had long-term exposure to vapor from mercury, which stabilized the wool in a process called felting after cutting hair from an animal pelt. The factory workers exposed to mercury vapor gave rise to the expression "mad as a hatter."

The most characteristic symptom, although it was seldom the first to appear, was mercurial tremors. The tremor often began in the fingers, but the eyelids, lips and tongue were also affected early. It progressed to legs and arms, and walking became difficult.

Severe neurological and kidney damage also occurred. After chronic exposure to the mercury vapor, the hatters tended to develop characteristic psychological traits such as pathological shyness and irritation.

The use of inorganic mercury in the form of mercuric nitrate used to treat small animal fur for felt hats began in the 17th century in France and then spread to England by the end of the century. By the Victorian era in the 1800s, the hatter's condition had become proverbial as expressed in the term "mad as a hatter." In 1898, a law in France was passed to protect hat makers from the risks of mercury exposure. In Britain, mercury poisoning had become a rarity in the 20th century after being curtailed for making hats. In the United States the disorder was thoroughly described in New Jersey in 1860, and the practice was continued until 1941 when WWII broke out. In 1878, 25 hat factories located around Newark, NJ were using mercury nitrate. The Essex county medical society revealed mercurial disease in 25 percent of 1,589 hatters. It was thought this was an underestimate, because hatters, afraid of losing their jobs, did not disclose symptoms.

Two-thirds of the recorded deaths of hatters in Newark and Orange were caused by pulmonary disease, most often in men under 30 years of age. Elevated death rates from tuberculosis persisted into the 20th century.

By 1934, the US Public Health Service estimated that 80 percent of American felt hat makers had mercurial tremors. The trade unions did not address the issue. It was the need for mercury in WWII that eventually brought an end to the use of mercuric nitrate in the United States.

Minamata Disease

Minamata disease is a disorder caused by methyl mercury poisoning first described by the inhabitants of Minamata Bay in Japan. Minamata Bay was contaminated with mercury originating from industrial waste. The mercury entered fish that was consumed by the locals of Minamata Bay. The disorder was characterized by peripheral sensory loss, tremors, dysarthia (difficulty in articulating words,) ataxia (an inability to coordinate voluntary muscular movement), and symptoms of hearing and visual loss.

Even the unborn child was at risk because methyl mercury readily crossed the placenta from mother to fetus, particularly into the developing brain. Children born with Minamata disease often had growth deficiencies, microcephaly, an abnormally small head, severe mental retardation, and they could be deaf or blind.

Minamata disease was not confined to Minamata Bay where the primary mercury source was contaminated fish. Other sources of maternal exposure has included flour made from grains treated with methyl mercury. At least 6,500 people from Iraq had been diagnosed. Another source was from meat of animals who were fed mercury tainted grain in New Mexico.

Acrodynia (Pink Disease)

Children under five years of age who were given teething powder ointment that contained mercurous chloride often became ill. The children suffered from irritability and sleeping disorders in a condition called acrodynia.

SOURCES OF MERCURY

There are many sources of mercury ranging from natural origins to medicinal sources. The natural sources result from degassing of the earth's crust, evaporation from natural bodies of water, and volcanic emissions. Between 30,000 to 150,000 tons are released into the atmosphere. Each year over 90 percent of atmospheric mercury is in the form of gaseous mercury vapor with less than 1 percent in particulate form. Mercury returns to the earth's surface in rain water in the form of Hg++ (Hg is the symbol for the element mercury) and CH3Hg+ (methyl mercury). Mercury vapor can remain in the atmosphere for several months and probably more than one year. The northern hemisphere has averaged 4 mg/m3, which is approximately double that of the southern hemisphere.

Mercury is also released into the air by human activities including fossil fuels (especially coal) combustion, from waste disposal, gold mining, cement production, metal smelting, refuse incineration, industrial application of metal, and cremations. Worldwide mining yields about 10,000 tons of mercury a year.

Human sources account for another 7,250 tons per year. In the diet, predatory fish account for the most mercury in the form of methyl mercury. Freshwater trout, pike, bass, tuna, swordfish, and shark account for the highest source of fish mercury.

USES OF MERCURY

In 2005, 20 metric tons of mercury were used domestically. Approximately 50 percent of it was used for chlorine-caustic soda manufacture. The remaining half was used for dental amalgams, thermometers, mercury electric switches, fluorescent lamps, and other electronic devices. Mercury is used for extraction of gold in the mining industry. It is also used as a cathode in electrolysis for sodium chlorides.

In the health care field, mercury can be found in laboratory and medical instruments. Mercury has been used as a therapeutic agent, and as a preservative in vaccines in the form of thimerasol (ethyl mercury). It has been used in teething powder, and in skin creams and soaps that lighten skin color. At one time they were used in laxatives that are now banned in the United States and Europe.

In agriculture, ethyl and methyl mercury have been used extensively in seed treatment. Most countries have banned this use. Phenyl mercury has been used as a fungicide and algaecide in paints and plastics. This type of mercury is unstable and slowly releases inorganic mercury.

Dental Amalgam

In 1991, the World Health Organization identified mercury in dental fillings (amalgam) as the largest source of mercury for humans. Sixteen years earlier, the WHO estimated that 3 percent of total mercury consumption was used for dental amalgams.

Amalgams have been used for over 150 years and accounted for 75 to 80 percent of all single tooth restorations. Each American dentist in private practice uses an average of 0.9 to 1.4 kg of mercury per year.

The conventional silver amalgam (silver filling) consists of a 1:1 mixture of mercury with an alloy consisting of silver (70 percent), tin (25 percent), copper (1- 6 percent), and zinc (0-2 percent). The amalgam begins to set within minutes of insertion and then needs to be carved to satisfactory anatomic form in that time period. Continued hardening of the amalgam restoration takes place over many months.

The modern method for preparation of amalgams involves mixing the alloy with mercury in a sealed capsule. In pediatric dentistry, a copper amalgam contains 60 to 70 percent mercury and 30 to 40 percent copper. The use of mercury in amalgams in Sweden was once known to be 5 - 7.5 tons per year compared to 90 - 100 tons in the United States. It is estimated that 170 - 180 kg of metallic mercury is released annually from a total of 50,000 cremations, with the mercury coming from the amalgam.

After insertion of amalgams into humans, there is a significant concentration increase of mercury in the urine and feces. In humans following amalgam insertion, there is a significant increase in urine mercury following insertion of four to five small occlusal fillings. After removal of amalgams, there is a higher peak of urine mercury several days after removal because of the mechanics of removing amalgams.

Mercury vapor released in the mouth from amalgams leads to an increase in mercury in body tissue. Studies have shown that the release of mercury vapor increases dramatically when the amalgam is stimulated by continuous chewing, reaching a plateau within ten minutes after the chewing subsides. It takes approximately

90 minutes for the mercury released to decline to the basal pre-chewing value.

Autopsy studies have correlated the number of amalgam fillings, or amalgam surfaces with mercury content in the brain and kidney. One study found amalgam-free subjects showed a mean mercury level in the kidney of 49 ng/g compared to amalgam subjects who had a level of 433 ng/g. In the brain of amalgam free subjects, the gray matter averaged 6.7 ng/g and the white matter averaged 3.8 ng/g. Those with amalgams had values of 15.2 ng/g in the gray matter and 11.2 ng/g for the white matter. Another study found females with amalgams had higher brain levels of mercury compared to men who had amalgams.

Studies on sheep found that mercury levels in maternal blood, fetal blood, and amniotic fluid reached a peak within 48 hours following amalgam placement in the mother and remained at that level for the duration of the 140 day study.

Studies have also shown that personnel working in dental offices have higher levels of mercury in their tissue.

THE CHEMISTRY OF INORGANIC MERCURY

Mercury is divided into inorganic and organic mercury. Organic mercury has a carbon atom attached to it. Because mercury (Hg) is the only element that exists as a liquid at room temperature, it was given the name "quick silver" by Aristotle over 2000 years ago. Inorganic mercury has three stable oxidative states.

1. Hg 0 is the metallic state and is called elemental mercury.
2. The loss of one electron gives rise to the mecurous state, Hg+.
3. The loss of two electrons results in the mercuric ion Hg++.

Elemental mercury has a very high vapor pressure, and it is in this state that mercury is given off by the dental amalgam. The vapor is inhaled into the lungs and then enters the bloodstream. Mercury vapor is more soluble in plasma, whole blood, and hemoglobin than in distilled water. Mercury from the dental amalgam is the dominant source of mercury exposure in terms of uptake and retention for the general population in industrialized countries. The uptake of mercury from the amalgam increases the tissue concentration in the brain, plasma, and kidney in proportion to the number of amalgams. The daily retention of mercury from amalgams is estimated at 3-17 ug. In the fetus, the mercury from the mother's amalgams enters the fetus's brain and kidney.

Inhalation of mercury vapor from the amalgam is the most common route of elemental mercury into the body. About 80 percent of inhaled mercury vapor is retained by the body. The retention occurs almost entirely in the alveoli of the lung. The dissolved mercury vapor is then soon oxidized to the mercuric form (Hg++) partly in the red blood cell and partly after diffusion into other tissue. The oxidation of mercury occurs under the influence of the enzyme catalase. On entering the blood component, mercury vapor dissolves in the plasma to the monatomic gas which is lipid soluble and highly diffusable. It rapidly leaves the plasma and enters the red blood cell. At this stage it readily crosses the blood-brain barrier. Once it is oxidized within brain cells to mercuric mercury, it becomes trapped in the brain. As a result, much of the mercury vapor deposition occurs in the brain. This accounts for the well-known effect of inhaled mercury vapor on the central nervous system. Subacute exposure to mercury vapor has given rise to psychotic reactions such as delirium, hallucinations, and even suicidal tendencies. Acute inhalation of mercury vapor can give rise to chest pain, coughing, dyspoena, and sometimes pneumonitis leading to death.

Mercury vapor in the blood is also transported to other body tissue where it is oxidized.

Organic Mercury

The other primary source of mercury exposure to humans is from diet, primarily from predator fish. This mercury is organic, called methyl mercury. In nature, methyl mercury is produced from inorganic mercury as a consequence of microbial activity. Methyl mercury found in fish originates from mercury in water and the bottom sediment.

The hazards of long-term intake of food containing methyl mercury are many, because it has a 90 percent absorption rate. The half time of methyl mercury is 70 days in the body, and it can accumulate in the brain. Chronic poisoning by methyl mercury results in degeneration and atrophy of the cerebral cortex causing parasthesia, ataxia, hearing loss, and visual impairment. Methyl mercury in pregnant women may result in inhibited development of the fetus, with psychomotor retardation of the born child.

Large tuna fish exceeding 60 kg have levels of up to 1 mg/kg of methyl mercury in their muscle. In polluted water, levels of methyl mercury in fish may exceed 10 mg/kg. There is a positive correlation between acidification of lakes and the presence of mercury. The average daily intake of fish for the Swedish population is estimated at 30g/day, resulting in a daily intake of methyl mercury between 1 and 20 ug.

Methyl mercury is efficiently absorbed from the gastrointestinal tract with about 95 percent of the isolated dose being absorbed. It is distributed to all parts of the body and readily crosses the blood-brain barrier and placenta barrier. Distribution is complete in two days and three days for the brain. Methyl mercury is slowly transformed in the body to inorganic mercury.

Methyl mercury passes through the blood brain barrier by entering the capillary endothelial cells as a complex with the amino acid cysteine attached to the thiol group of cysteine. Methyl mercury cysteine is being formed from bile of the glutathione complex. Methyl mercury is then exported from liver cells into the plasma with glutathione.

Molecular Mechanism of Toxicity

The nervous system is the most sensitive tissue to mercury exposure. The mode of toxic actions of mercury on neural membranes is by its effect on protein biosynthesis, microtubules, nucleic acid structure, and oxidative stress on cells.

Neurological symptoms that occur after years of exposure to mercury vapors include mental impairment and dementia. Mercury in any chemical form will denature protein by deactivating enzymes leading to disruption of tissue. Neuromuscular junctions are a major target of mercury compounds, with end plate potential being suppressed by methyl mercury.

Methyl mercury stimulates the release of calcium (CA++) from nerve terminal mitochondria. As a result, resting membrane and action potentials are decreased.

Free radicals cause oxidative injury to cells in the pathogenesis of methyl mercury neurotoxicty. The combination of changes in intracellular calcium homeostasis and disruption of sulfhydryl status initiates activation of phospholipase, proteases, and endonuclease, which all lead to neural injury. Inorganic mercury has been found to cause either autoimmune responses or immune suppression.

DNA is a target of free mercury radicals. Mercury induces DNA single strand breaks confirmed by cytochrome C reduction. The mercuric ion effectively inhibits DNA repair. Methyl mercury has been shown to be more potent in causing DNA strand breaks compared to inorganic mercury.

Methyl mercury affects the synthesis of individual proteins. Mercury affects protein synthesis of actin and tubulin, which are cytoskeletal proteins. This inhibits tubulin biosynthesis through messenger RNA synthesis. Methyl mercury also targets micro tubules, which depresses their biosynthesis and increases the cellular tubulin pool. Tubulin messenger RNA is suppressed by methyl mercury.

The central nervous system is the most critical organ for mercury toxicity. Elemental mercury vapor and methyl mercury both cause nerve damage. The nerve membrane structure is the first target of mercury in conjunction with lipid peroxidation. Methyl mercury can cause changes to alpha-norepinephrine receptors.

The developing brain has been considered to be the most sensitive to methyl mercury. In utero exposure at low-dose levels can cause psychological and behavioral problems in the offspring. Mercury accumulates in the cerebral cortex, especially in the occipital and parietal cortical areas. Monkey studies show mercury accumulates in neurons and astrocytes as well as in the pyramidal cells of the hippocampus. In the cerebellum, the highest amount of mercury is found in both neurons and glial cells. Even three years following cessation of mercury exposure, mercury is still found in the monkey cerebellum. It also accumulates in the retina of the eye.

Methyl mercury reacts with DNA and RNA, and binds to sulfhydryl groups causing a reduction of DNA synthesis. It disrupts protein synthesis, and rat studies found a 50 percent inhibition of protein synthesis in synaptasomes in the rat brain cortex. There is a marked inhibition of mitochondrial biogenesis. Methyl mercury inhibits the polymerization of tubulin by 60 percent. Tubulin is the basic element in the microtubules of the brain cells. Micromolar concentrations of methyl mercury have been found to disrupt calcium homeostasis causing increased intracellular calcium concentrations.

SYMPTOMS OF MERCURY TOXICITY

Symptoms following mercury vapor exposure are many. This results from mercury strongly binding with sulfhydryl groups which affects enzymes. It interacts with nerve receptors, ion channels, and intra-cellular signals. The central nervous system is the most sensitive organ to mercury. Often it leads to genetic polymorphism. Symptoms, which may vary following chronic exposure, include weakness, fatigue, memory loss, insomnia, depression, delirium, hallucinations, anorexia, loss of weight, gastrointestinal disturbances, and tremor. Chronic mercury poisoning in cells is called erethism or micromercuralism. Other symptoms include cognitive deficits, poor motor coordination, poor problem solving, and poor comprehension.

Occupational exposure to mercury causing permanent damage has been found decades after exposure has ceased. Animal studies have shown exposure to mercury vapor gives rise to abortion, increased neonatal death, developmental disorders in the brain, and reduced learning. For many years the mercury preservative thimerosal was used in vaccines and evidence suggested it may have been involved with autism. One study of 917 children (7 years old) whose mothers had been exposed to methyl mercury from consuming pilot whales, found attention deficit disorders and memory deficits.

Memory problems have been a very common symptom in a number of studies in people who were exposed to mercury vapor. One study of 185 workers exposed to metallic mercury vapor found symptoms of memory disturbance, depression, fatigue, and irritability compared to a control. Another study in 1984 examined

36 choralkali workers, who were exposed to mercury, and were compared to a control group. The workers exhibited decreased memory and verbal intelligence. A study of short term memory of mercury-exposed workers found memory deficit correlated with urine mercury and to the amount of mercury exposure.

A number of studies have associated menstrual disturbances in women exposed to elemental mercury vapor in the workplace. They were compared to non-exposed workers.

One study found the percentage of women having normal menstrual cycles at the commencement of employment to be similar to the control. The mercury-exposed group then developed changes in their menstrual cycle. Another study found mercury exposed workers experienced more spontaneous abortion (17%) compared to the control (5%). There was more toxemia of pregnancy, 35% to 2%.

For humans, a large proportion of the population have dental amalgam fillings that continuously give off mercury vapor. Long-term exposure to mercury concentrations at levels of 0.1 mg/m3 in air or higher can manifest mercuralism that includes changes in behavior and possible tremor. Animal studies with mercury vapor found the same toxic signs and brain pathology as methyl mercury.

Ethyl mercury is found in thimerosal preservative used in medications. Ethyl mercury is a neural development toxicant, with clinical symptoms similar to methyl mercury. One study found a decline in autism rates after the ban in child vaccines.

EXCRETION OF MERCURY

Animal studies have shown that after a single exposure to mercury vapor, 10 times more mercury is retained in the brain compared to intravenous injection of the same dose of mercuric mercury. The mercury from mercury vapor has an affinity to certain types of cells. The elimination of elemental mercury vapor occurs by excretion in the form of mercuric mercury because the mercury has been oxidized to the mercuric form. The bulk of accumulated mercury in the body, about 80 percent, is excreted with a biological half time of 60 days. In the brain, the half time may exceed several years to several decades.

The main route of methyl mercury elimination is through the liver into the bile and then the feces, and also through the kidney in urine. Mercury has an affinity for the kidney after being absorbed by the blood. There are two types of kidney injuries resulting from mercury exposure, including tubular damage and glomerular injury. The earliest signs of inorganic mercury toxicity on rats occurred in the kidney proximal tubule. However, the major route of methyl mercury excretion is through the feces, about 90 percent of it. Methyl mercury undergoes biotransformation to inorganic mercury by demethylation in the body accomplished by OH- radicals. Biotransformation in the brain occurs at a slow rate. About 80 percent of mercury in breast milk is in organic form.

This brief overview of mercury toxicity has demonstrated how mercury can affect health. The largest exposure to mercury comes from the dental amalgam. Could it be causing physical and mental health problems? The answer is yes, as you will discover in the next few chapters. It may also be this source of mercury that causes Alzheimer's disease.

CHAPTER THREE

The Dental Amalgam

The Greatest Source of Mercury Exposure

The dental amalgam has historically been the most common filling used for dental cavities. Often called a silver filling, 50 percent of the amalgam is comprised of mercury accounting for 75 to 80 percent of all single tooth restorations in 1976. Since the current amalgam mercury controversy began in the 1980s, many dentists have reduced or eliminated the use of amalgams. Silver, tin, copper, and zinc comprise the other 50 percent. The amalgam is continuously releasing mercury vapor that enters the lung and then the blood. From there it is transported to the brain and other body tissue. Once it enters the brain in the non-ionized state, it becomes ionized and entrapped within the brain, sometimes for decades. Evidence will be provided that it can cause a variety of mental health problems.

Much of the next three chapters is based on my Ph.D. dissertation at Colorado State University and my research with the Rocky Mountain Research Institute. Following my departure from CSU, I had received a number of research grants from the Wallace Research Foundation headed by H.B. Wallace, the son of Wallace, the former vice president of the United States under

President Franklin Roosevelt. The research investigated the health effects of dental amalgam mercury. While at CSU, the University of Colorado dental school helped fund my research.

In the 1980s and 1990s, Dr Hal Huggins, a dentist from Colorado Springs discovered that when he removed dental amalgams from his patients, their health often improved. Dr. Huggins was the person most responsible for raising consciousness about health effects from mercury poisoning. Of course, the dental profession did not want to know about this and eventually the Colorado Dental Association was influential in causing the Colorado Board of Dental Examiners to revoke Dr. Huggin's dental license. They play hardball as I soon found out.

In the 1980s, Dr. Huggins began raising consciousness about amalgam mercury poisoning. I had completed a master's degree in 1981 in nutrition at Colorado State University (CSU). My thesis was on the relationship of myopia (nearsightedness) and nutrition as I was still practicing optometry at my office across from CSU. Part of my study evaluated hair minerals of myopes (nearsightedness) compared to nonmyopes. The most significant finding of the whole study was that myopes had significantly lower hair mercury that I could not explain.

About that time, I began hearing of Dr. Huggins' claim about dental amalgam mercury toxicity. I then wondered if there might be a connection between dental amalgam mercury and myopia. While still practicing optometry, I decided to enter the PhD program in physiology at CSU to do a dissertation on the health effects of dental amalgam. My major advisor was Dr. David Robertshaw, the head of the physiology department. My research found that mercury from dental amalgams was causing all types of health problems, especially mental health issues.

Shortly before I finished my dissertation, Dr. Robertshaw accepted a position to be Department Head of Physiology at Cornell University. I finished all the course work and research that was requested by my committee. Dr. Robertshaw flew in to Colorado from New York for my Ph.D. dissertation defense. To my utter surprise, two members of my committee reneged on our agreement and wanted more research done and promptly resigned from my committee. I am sure they did not want to be associated with the controversy about dental amalgam toxicity and jeopardize their careers. I was devastated.

The new department head of physiology, Rupert Aman, did not want the research done. He put up many roadblocks for me such as requiring more committee members than required by the department. I was able to get a local neurologist to be the extra committee member after replacing the two committee members that left. I then received a grant to determine if mercury was related to multiple sclerosis. In the meantime, Dr. Robertshaw encouraged me to publish the original research in my dissertation. Dr. Aman did not want me to publish the research findings without being approved by the department. I felt that CSU was going to suppress my original research, and I submitted a paper to a journal and did not go through Dr. Aman. I had given a copy of the paper to another faculty member in physiology. Dr. Aman found out that I submitted a paper and promptly dismissed me from the department.

I hired an attorney and brought a lawsuit against CSU for illegally dismissing me from the program. The judge ruled in my favor, but my attorney had missed the statute of limitations of one year, and CSU would not let me back in. Essentially CSU owes me a Ph.D. degree.

After my dismissal, I received a number of research grants from the Wallace Genetic Foundation which later changed its name to the Wallace Research Foundation. I have had 20 scientific papers published regarding my research on dental mercury toxicity that will be discussed.

The ordeal taught me that academic science is very political. Their purpose I feel is to maintain status quo. There is so much bias for not getting involved in a controversial subject and alternative health. Academic scientists would rather preserve status quo than get involved in a controversy that might help mankind even if there is good science underlying it. I call it corruption. Government research grants will not be issued because of this extreme bias no matter how much evidence supports the hypothesis. The same with many scientific journals and their reluctance to publish important data. After I completed my research, funded from a private foundation, I concluded that dental amalgam mercury is one of biggest health hazards in the world today, as you will see.

In the 1990s. I received a sizeable grant to study the relationship of Alzheimer's disease and dental amalgam mercury. Tom Warren had written a book entitled *Beating Alzheimer's* after being diagnosed with AD, and then having his dental amalgams and teeth removed. He recovered from this dreaded disease and wrote a book. We asked the Alzheimer's Association to help recruit Alzheimer's patients for the study. They refused by saying that mercury does not cause Alzheimer's disease. We tried to recruit AD patients through newspaper ads, and realized that Alzheimer subjects do not read the newspaper. We then used the grant money to study schizophrenia, and as you will see in another chapter, schizophrenia improved following amalgam removal.

My research was done under the auspices of the Rocky Mountain Research Institute, Inc., whose president and co-founder was Maury Albertson, Ph.D., the former director of research at CSU who co-founded the Peace Corps and was nominated for a Nobel Peace Prize. It was Maury and associates who designed the Peace Corps that President Kennedy adopted.

History of Amalgams (The Amalgam Wars)

Amalgams have not always been accepted by the dental profession as witnessed by three "Amalgam Wars." In 1839, the amalgam was introduced in the United States by the Crowcour brothers who came to New York from Great Britain. The two dentists called them silver fillings and advertised them as a substitute for gold because they were less expensive. Then began the first amalgam war by American dentists who began a relentless crusade against the Crowcours and forced them to quit. The amalgam was denounced as poor filling material and was labeled as a cause of mercury poisoning. The use of amalgam was unanimously banned by several American dental associations. Eleven members were suspended in 1848 from the dental society in New York for malpractice because they used amalgams. Many of the resolutions prohibiting the use of amalgams were withdrawn by the American Society of Dental Surgeons in 1855 in order to put an end to the many internal conflicts within the society.

Dr. J. F. Flagg organized a movement on behalf of the amalgam in the late 1870s. Research began on the technical properties of the amalgam after it became recognized that the amalgam was a valuable filling material. Researchers found that silver led to expansion of the amalgam, while tin led to shrinkage. In the early 1900s, it was agreed that the perfection of dental amalgam had been obtained. The idea that the amalgam was toxic was completely forgotten. (Remember, the first case of Alzheimer's disease was identified in 1906.)

Not until 1926 did the controversy arise again beginning the second amalgam war. Professor Alfred Stock, a German chemist and pharmacologist, began a campaign against mercury. He published a paper regarding his own personal experience with mercury toxicity, describing how mercury destroyed a major part of his life. He had been poisoned by mercury vapor from his lab as well as from his dental amalgams. For the next two decades, Stock researched mercury toxicity based on clinical observation of himself, animals, and in vitro. Stock concluded that "dentistry should completely avoid the use of amalgams for fillings, at least not use it when possible." He said that "it will likely be found that the thoughtless introduction of amalgams as a filling material for teeth was a severe sin against humanity." After the press reported Stock's concern, the controversy continued until WWII, and it was again forgotten.

In the early 1980s, the third amalgam war began about the same time in the United States and Sweden. Swedish neurobiologist Dr. Mats Hanson, from the University of Lund, reopened the fight against the amalgam. Through research, press releases, scientific articles, literature reviews, and legal means, he tried to eliminate amalgams in Sweden without much success. He increased public awareness immensely. Dr. Hanson believed amalgams had "been the biggest poisoning catastrophe of the century." Other Swedish scientists such as Patrick Stortebecker, M.D., Ph.D., Jaro Pleva, Ph.D., and dentist Magnus Nylander have substantiated many of Hanson's claims through good research. Sweden has now banned amalgams, and Alzheimer's disease is on the decline in Sweden.

Hal Huggins, DDS ch9, led the fight against dental amalgam in the United States. Through clinical evidence, Huggins had anecdotally substantiated in amalgam patients, many of the biological and clinical symptoms reported in scientific literature.

Through his nationwide workshops and media exposure, Huggins had stimulated many of the nation's dentists to discontinue the use of dental amalgam. He had been the alarmist that has stimulated some scientists to research this important health problem.

The Amalgam

In 1991, the World Health Organization identified mercury in dental amalgam as the largest source of mercury for humans. Sixteen years earlier, the WHO estimated that 3 percent of total mercury consumption was used for dental amalgams. Amalgams have been used for over 150 years and accounted for 75 to 80 percent of all single tooth restorations. At that time, each American dentist in private practice used an average of 0.9 to 1.4 kg of mercury per year.

The conventional silver amalgam consists of a 1:1 mixture of mercury (50%) with an alloy consisting of silver (28%), tin (14%), copper (0.3%), and zinc (8%). The amalgam begins to set within minutes of insertion and then needs to be carved to satisfactory anatomic form in that time period. Continued hardening of the amalgam restoration takes place over many months.

The modern method for preparation of amalgams involves mixing the alloy with mercury in a sealed capsule. In pediatric dentistry, a copper amalgam contains 60 to 70 percent mercury and 30 to 40 percent copper.

The use of mercury in amalgams in Sweden was once 5 - 7.5 tons per year compared to 90 - 100 tons in the United States. It is also estimated that 170 - 180 kg of metallic mercury is released annually from a total of 50,000 cremations, with the mercury coming from the amalgam.

After insertion of amalgams in humans, there is a significant concentration of mercury in the urine and feces. In humans following amalgam insertion, there is a significant increase in urine mercury following insertion of four to five small occlusal fillings. After removal of amalgams, there is a higher peak of urine mercury several days after removal because of the mechanics of removing amalgams.

Mercury vapor released in the mouth from amalgams leads to an increase in mercury in body tissue. Studies have shown that the release of mercury vapor increases dramatically when the amalgam is stimulated by continuous chewing, reaching a plateau within ten minutes after the chewing subsides. It takes approximately 90 minutes for the mercury released to decline to the basal pre- chewing value.

Autopsy studies have correlated the number of amalgam fillings, or amalgam surfaces with mercury content in the brain and kidney. One study found amalgam-free subjects had a mean mercury level in the kidney of 49 ng/g compared to amalgam subjects who had a level of 433 ng/g. In the autopsied brain of amalgam free subjects, the gray matter averaged 6.7 ng/g and the white matter averaged 3.8 ng/g. Those with amalgams had values of 15.2 ng/g in the gray matter and 11.2 ng/g for the white matter. Another study found females had higher brain levels of mercury compared to men who had amalgams.

Studies on sheep found that mercury levels in maternal blood, fetal blood, and amniotic fluid reached a peak within 48 hours following amalgam placement in the mother and remained at that level for the duration of the 140 day study.

Studies have also shown that personnel working in dental offices have higher levels of mercury in their tissue.

At one time, dental practices throughout the world used about 300 tons of mercury annually. In the United States, the dental profession utilizes about 250,000 pounds, with each dentist averaging 2.5 pounds. With the amalgam being comprised of 50 percent mercury, the amalgam is an alloy that continuously gives off mercury vapor. Insidious long term exposure to mercury can produce health problems and symptoms after many years, while the cause can be difficult to establish.

Once the elemental mercury vapor is inhaled, it can be oxidized in the blood and become a mercuric ion which forms bonds with certain tissue ligands such as thiol groups in protein. Internal bacteria such as streptococcus mutans is known to methylate amalgam mercury, which is organic mercury in the most toxic form that readily crosses biological membranes.

Newer fillings release more mercury than older fillings, and chewing causes more mercury to be released. After chewing, there is an average of 15.5 times more mercury vapor compared to prechewing levels. Mercury amounts of up to 87.5 mcg/m3 of air have been measured in some subjects. Vimy and Lorscheider found subjects with 12 or more occlusal amalgam surfaces received an average daily dose of 29 mcg of mercury, and those with 1 to 4 amalgams received an average of 9 mcg. These levels exceed the safety threshold of many countries accepted standard of environmental mercury exposure. Radic, et al, analyzed an outer corroded zone of an amalgam that had been depleted of mercury. It was calculated that a mouth with many fillings would release up to 560 mg of mercury over several years. If the loss was over 10 years, the daily dose would be 150 mcg.

Both chewing and tooth brushing enhance the release of mercury. Evaporation of mercury is intensified rapidly during intake of warm food and drink. Mercury compounds are easily dissolved by weak acids and sodium chloride solutions. Weak acids from citrus foods will do the same. Streptococcus mutans, known to cause dental caries, also enhances corrosion of dental amalgam.

Mercury can also be released inward, as amalgam mercury has been found in neighboring enamel. Ionized mercury can be taken up by tissue and nerves, and it can migrate from root to jawbone when an amalgam has been in constant contact with gold. Levels can reach more than 1,200 mcg per gram of tissue, but only a few hundred mcg of mercury per gram of tissue when the amalgam is not in contact with gold.

When mercury is taken up by nerve endings, it is transported toward the central nervous system by retrograde axonal transport, the same way lead is transported.

Amalgam mercury can form soluble mercuric chloride that easily disassociates into mercuric ions. In the intestine, they are easily absorbed into the mucous tissue.

Daily mercury concentration of 10 to 20 mcg/m3 in the work room was found to produce both physical and mental symptoms among the majority of employees. In both Europe and the United States, the maximum allowable concentration is 50 mcg/m3 in air. In the Soviet Union, the standard is set at 10 mcg/m3 of air. This concentration is based on a daily exposure of eight hours during a five-day work week.

The mercury released from dental amalgam is predominantly in the from of mercury vapor that can enter the body in several ways. It can enter saliva which passes to the gastrointestinal tract, which is either eliminated by feces or absorbed into the blood. Mercury vapor can be inhaled into the lungs, and an average of 75 to 80 percent can be absorbed by the aveoli and pass into the blood circulation and then the brain. The vapors of mercury that settle down in the oronasal region may be absorbed and pass directly into the valveless cranial venous system, and in this way enter the brain. Mercury can be transported to the olfactory bulbs on the skull base and then into the brain. Stock found a high content of mercury in the olfactory lobe and frontal brain. Dr. Nylander confirmed this and also found higher levels in the pituitary gland. There is a direct correlation between the number of occlusal surfaces of amalgams and the mercury levels in the brain.

The EU has prohibited amalgam use in children under 15 and for pregnant or breast feeding women.

Corrosion and the Oral Galvanic Cell

The way mercury is released from dental amalgam is through corrosion. Corrosion of amalgam occurs because of the formation of galvanic cells similar to a dry cell battery. A galvanic cell requires an electrolyte which is the saliva in the mouth. Two kinds of galvanic cells exist within the oral cavity, a bimetallic cell and differential aeration cell.

In the bimetallic cell, the more noble metal will be the cathode, which is often gold. The less noble metal is the anode, and this is usually the amalgam which will dissolve.

In the differential aeration cell, there are differences in oxygen concentration around a single restoration which will form a galvanic cell. In crevices, which occur almost always in amalgams, the oxygen is depleted and the electrolyte is enriched with metal ions and other ions such as chloride. The cathode of a differential aeration cell will form on the part of the metal surface, which is in contact with an electrolyte of high oxygen concentration, and the anode will form in an area of less oxygen.

The magnitude of the corrosion current depends on the conductivity of the electrolyte saliva or the oxygen partial pressure. Foods such as vinegar and salt will increase conductivity as does low PH. Increased current releases more mercury. This is galvanic corrosion. Mercury is enriched on the amalgam surface and evaporates as elemental mercury vapor. With time, the amalgam becomes porous, and the increased surface area enhances corrosion. Amalgam has been found to corrode faster than any other dental restorative material tested. Many agents can chemically corrode an amalgam. Sulfur is considered to be the important dietary element responsible for corrosion of amalgam. Electrochemical corrosion is the other type of oral corrosion.

Oral Galvanic Current

When gold and amalgam or other dissimilar metals are placed side by side in the mouth, an electrical current is established, producing currents of up to 25 micro amperes and voltages up to 1000 millivolts or 1 volt. Metal restorations are in electrolytic contact through the saliva and bone fluid. Electric currents will always form a closed circuit resulting in a current from an amalgam, especially if there is gold next to it.

Currents greater than 10 micro amperes have been measured in new single amalgams. A current of 100 micro amperes will release at least 1.5 mg of mercury each day. These levels subside after several months. In root filled teeth, the resistance can be as low as 1.5 kilo ohms resulting in a high current and easy migration of mercury ions into tissue and circulation.

Evidence has been presented how mercury leaves the amalgam and gets into the brain and body tissue. The following three chapters will provide evidence how mercury from amalgams may be affecting health.

CHAPTER FOUR

The Relationship Between Mercury from Dental Amalgam and Health

It's Worse Than You Think

It was my major advisor, Dr. David Robertshaw, the Department Head of Physiology at Colorado State University, who suggested we compare health differences between people with dental amalgams and those without. I asked myself, why didn't I think of that, as it sure made sense. We asked the University of Colorado dental school to participate in the research, and they reluctantly agreed to help fund the research.

The volunteers consisted of 51 individuals without dental amalgams and 50 subjects with amalgams. The amalgam group of 30 females and 20 males was compared to 30 females and 21 males in the group without amalgams. Seventy of the volunteers were recruited through newspaper ads, and 23 subjects volunteered from various physiology labs. The amalgam group averaged 23.3 years in age and the non-amalgam group averaged 22.4 years. There was no significant difference in weight or height of the subjects. The males had an average of 10.1 amalgams and the females averaged 9.8 amalgams. Fourteen of the non-amalgam subjects had heard of the amalgam mercury controversy, and 18 in the amalgam group were aware of it.

The first stage of the research involved a detailed health questionnaire, a comprehensive eye exam, cardiovascular evaluation, and a blood profile. The second phase of the research studied 87 subjects who had their amalgams removed. The third stage of the research involved determining if mercury caused free- oxy radicals, the cause of many health problems. Statistical analysis was done by the Colorado State University Statistics Laboratory. Most of the findings that had a p value (probability) of 0.10 or better will be discussed. A p value of 0.10 means the probability of the finding happening by chance was 1 in 10. A p value of 0.05 means it could have happened by chance 5 percent of the time or 1 in 20.

General Health

The study provided evidence that subjects with dental amalgams were less healthy than individuals without amalgams. The amalgam volunteers had 45 percent more health symptoms per subject than the non-amalgam group. Out of 125 health history questions, the amalgam subjects averaged 17.12 symptoms of ill health compared to 11.78 symptoms per subject in the non-amalgam faction. This was significant at the 0.0001 level. The amalgam group had 98 symptoms out of 125 symptoms in which they out numbered the non-amalgam group with 14 of the symptoms being significantly different. The non-amalgam volunteers had 19 symptoms in which they outnumbered the amalgam subjects with three being significantly different. The health categories in which the amalgam group had significantly more symptoms included digestion (p=0.04), endocrine (p=0.019), mental and emotional (p=0.0035), annoying symptoms (p=0.068), general disease (p=0.025), and dental (p=0.007). In the remaining six categories of skin, cardiovascular, nervous, blood, allergies, and the eye, it was found that the subjects with amalgams had more symptoms, but the differences were not at the significant level of p=0.10. Most of the significant symptoms exhibited by the amalgam group could be explained by mercury toxicity.

Mental and Emotional

Mental and emotional problems were the areas of health most significantly affected in the amalgam group. They had significantly more symptoms of sudden anger (amalgams = 10, non-amalgams = 1, p=0.045), more depression (amalgam = 18, non-amalgam = 7, p=0.008), more irritability (amalgam = 19, non-amalgam = 9, p=0.018), and they were less happy (p=0.047). The subjects in the amalgam group had 143 percent more mental health issues than the non-amalgam group (p=0.007). "Mad hatter" was the term used to describe the mental illness exhibited by workers using mercury in the felt hat factories in the 19th century. A considerable number of volunteers with amalgams had symptoms similar to the mad hatter syndrome. These findings strongly suggest that mercury from amalgams is already affecting health in young adults, especially mental health. Once inside the brain, mercury becomes ionized and entrapped within the brain for up to several decades.

The study found other mental problems were greater in the amalgam group but not at the significant level, including lack of self-confidence, trouble with decisions, lack of attention, nervousness, nightmares, inability to concentrate, suicidal tendencies, and anxiety. When asked to describe their reading comprehension, 41 non-amalgam subjects responded "good" as compared to 35 amalgam volunteers (p=0.04) suggesting that mercury from amalgam may be affecting learning.

Immunity

The amalgam group had significantly more annual colds and respiratory infections, n = 17, than the non-amalgam group, n = 9, (p=0.046). Nine of the non-amalgam subjects reported they had a history of hay fever compared to 16 in the amalgam group (p=0.026). The blood tests that will be discussed later showed the amalgam group had a lower number of T lymphocytes, and their immunoglobulins correlated significantly with the number of amalgams suggesting the immune system is affected.

Nervous System

One of the classical symptoms of mercury toxicity is tremors. The amalgam group reported significantly more tremors (n = 17) compared to the non-amalgam group (n = 4) (p=0.002). The development of involuntary movement is an early manifestation of central nervous system involvement from mercury.

Endocrine

The amalgam subjects had 60 percent more symptoms (p=0.019) in the 12 health questions involving the endocrine system. Menstrual problems were the most symptomatic area with 16 amalgam females having them compared to 10 in the non-amalgam group (p=0.09). (Mercuric chloride has been found to suppress progesterone in hamsters as well as their estrous cycle.)

Gastrointestinal

The amalgam group reported 77 percent more symptoms in the digestion category (p=0.04) with heartburn being the only significant symptom (amalgam = 10, non-amalgam = 4, p=0.07). Amalgam subjects also reported more indigestion, ulcers, diarrhea, and a bloated feeling. Gastrointestinal problem have long been associated with chronic inorganic mercury vapor poisoning.

Oral Cavity Health

In my studies, I found that dental amalgam removal significantly improved oral cavity health, findings which were published in *Annals of Dentistry* (winter 1990). When subjects with amalgams were compared to subjects without, they had more symptoms of foul breath, bleeding gums, grinding teeth, metallic taste, and periodontal disease. Another study involved 87 subjects who had their amalgams removed. They reported a total of 125 oral cavity symptoms before amalgam removal. Following amalgam removal, 32% were eliminated and 54% improved. Eighty-six percent of the periodontal symptoms were improved or eliminated.

We also compared oral cavity health of 47 multiple sclerosis patients with amalgams averaging 41.4 years in age to 47 multiple sclerosis patients who had their amalgams removed who averaged 44.3 years in age. The finding suggested that amalgam mercury may play a role in the etiology of oral cavity health. Subjects without amalgams averaged 0.72 oral cavity symptoms per subject compared to 1.28 symptoms per subject for those with amalgams during the past 12 months. Subjects who had their amalgams removed reported that 33 percent of their oral cavity symptoms were eliminated and 39 percent of their symptoms improved.

Lifestyle

More amalgam subjects smoked (n = 6) compared to the non-amalgam volunteers (n = 1), p=0.08. A Swedish study found a direct correlation between number of amalgams and amount of smoking when compared to a control without amalgams. Mercury is known to inhibit the uptake of dopamine and serotonin at synapses. Mercury affects norepinephrine in brain synaptosomes and also inhibits acetylcholine. On the other hand, nicotine has the opposite effect on these neurotransmitters. It can increase levels of dopamine, serotonin, acetylcholine, epinephrine, and norepinephrine. The amalgam subjects suffered significantly more from depression, irritability, and anger perhaps caused by the affected neurotransmitters. Smoking may be an anecdote to overcome the depressed neurotransmitters.

The amalgam females took the birth control pill (n = 11) significantly more than the non-amalgam group (n = 4), p=0.019. Besides birth control reasons, birth control pills are used for therapy for menstrual problems that more amalgam subjects suffered from.

The health questionnaire suggested there was something going on with health in subjects with amalgams, as they suffered from 44 percent more symptoms than non-amalgam subjects. Could it be mercury in the fillings causing these symptoms? During this time period, the media was raising consciousness about possible mercury poisoning from dental amalgams. The news program *60 Minutes* carried a segment on the controversy and had asked me to send a copy of my Ph.D. dissertation for them to review. Once people realized they may be poisoned by their dental amalgams, they began to have their amalgams removed. This was the next step in the research, to determine if health improved following amalgam removal.

Health Effects After Dental Amalgam Removal

Many types of health problems are associated with mercury poisoning. If mercury from dental amalgam was causing health symptoms, one would expect the symptoms to improve after amalgam removal. Anecdotal evidence does provide confirmation that symptoms do improve as mercury-free dentists were finding out. Several papers in the scientific literature also suggested this. Anecdotal evidence does not involve a control group, but still raises the question there may be a placebo affect.

Dr. H. Raue, a German eye physician, had done much work with patients poisoned by mercury from dental amalgam. He based much of his diagnosis on the electrical current between amalgams. In two years, he measured 100 patients who had elevated oral galvanic current measurements over 5 microamperes, with symptoms of micromecurialism that disappeared following amalgam removal. Many of Raue's patients had symptoms that had been resistant to traditional therapies by other physicians but disappeared after amalgam removal. Symptoms that disappeared after removal included migraine headaches, nausea, frequent blackouts, headaches, poor equilibrium, dizziness, confusion, poor memory, suicidal tendencies, sleep disorders, tinnitus, cardiovascular problems, intestinal disorders, depression, psychiatric disturbances, and allergies. Most of these symptoms disappeared on the 100 subjects that Raue studied.

One of the most informative articles written about amalgam toxicity was written by Jara Pleva, Ph.D., a corrosion scientist in Sweden. For over 20 years, Pleva had a multitude of health problems that physicians could not cure. It was only after he discovered the corrosion of his amalgams that he began to suspect mercury poisoning. He also had a dental cavity that was root filled with amalgam by drilling through a gold filling. After the amalgams were removed, all the symptoms quickly disappeared within three months, except for facial paralysis and *arcus senilis* on the cornea. His mental abilities and memory recovered more slowly than the somatic functions. The mental health symptoms that improved following Dr. Pleva's amalgam removal were anxiety, irritability, difficulty in controlling behavior, indecision, loss of interest in life, feeling old, resistance to intellectual work, increased need for sleep, and stress. The somatic symptoms that improved were joint pain, headaches, gastrointestinal irritations, irregular heart beat, pleurisy, and inflammation to name a few.

The literature confirmed people's health did improve following amalgam removal, so the next step in our research was to study people who had their amalgams removed. Dr. Ed Harrison, a dentist in Utah, provided nearly 300 names of patients from whom he had removed amalgam fillings. The questionnaire asked his patients to describe their health status, health symptoms, and lifestyle for the year before amalgam removal and the time after removal. Eighty-six people volunteered who had their mercury fillings removed,

Results

Part I

The subjects were asked to rate their status on a scale of 1 to 10, before and after amalgam removal.

A. Rate your tolerance to stress.
    Before: 4.85
    After: 7.17
    Percentage Improvement: 47.8%

B. Rate the amount of stress you are under.
    Before: 6.85
    After: 6.44
    Percentage: 6.7% less stress after removal

C. Rate your emotional level.
    Before: 6.22
    After: 5.04
    Percentage Improvement: 19%

D. Rate your overall health.
    Before: 6.09
    After: 7.69
    Percentage Improvement: 26.2%
E. Rate your overall happiness.
    Before: 6.27
    After: 7.92
    Percentage Improvement: 26.4%
F. Rate your overall peace of mind.
    Before: 6.15
    After: 7.85
    Percent Improvement: 27.6 %
G. Rate your overall reading comprehension.
    Before: 6.78
    After: 7.89
    Percent Improvement: 16.3 %

Part II

The health questionnaire had 12 categories of questions with a number of symptoms in each category. The 86 subjects were asked if the symptom was eliminated, improved, no change, or worse. The three categories that showed the most improvement and elimination of symptoms were mental health (82%), cardiovascular (83%), and dental (86%). The following table is a summary of the results of each category.

Category	Total	No Change	Eliminated	Improved	Worse	Eliminated + Improved
1. Skin	87	17 (22%)	24 (28%)	40 (46%)	4 (5%)	74%
2. Cardiovascular	51	8 (16%)	11 (22%)	31 (61%)	1 (2%)	83%
3. Nervous	9	19 (19%)	19 (19%)	57 (58%)	3 (3%)	77%
4. Digestion	115	25 (22%)	16 (14%)	70 (61%)	3 (3%)	76%
5. Blood	7	0	5 (71%)	2 (29%)	0	100%

6. Endocrine	106	44 (44%)	9 (8%)	51 (48%)	2 (2%)	56%
7. Mental	409	61 (15%)	62 (15%)	275 (67%)	10 (2%)	82%
8. Annoying Symptom	479	171 (36%)	71 (15%)	226 (47%)	16 (3%)	62%
9. Allergies	92	34 (37%)	2 (2%)	56 (61%)	0	63%
10. Diseases	85	16 (19%)	14 (6%)	50 (59%)	5 (6%)	65%
11. Eye	172	92 (53%)	13 (8%)	56 (33%)	11 (6%)	41%
12. Dental	125	13 (10%)	40 (32%)	67 (54%)	10 (4%)	86%
Total Symptoms	1826	502 (27%)	286 (16%)	981 (54%)	60 (3%)	70%

Mental Health Questionnaire (86 Subjects)

Symptom	Total	No Change	Eliminated	Improved	Worse	Eliminated + Omproved
1. Sudden Anger	31	3 (10%)	5 (16%)	23 (74)	0	90%
2. Depression	38	4 (11%)	4 (11%)	28 (73%)	2 (5%)	84%
3. Wish Dead	17	0%	7 (41%)	10 (59)%	0%	100%
4. Irritability	34	5 (15%)	4 (12%)	25 (73%)	0%	85%
5. Suicidal	10	0%	3 (30%)	7 (70%)	0%	100%
6. Divorced	5	1 (20%)	0%	4 (80%)	0%	80%
7. Frequent Anxiety	35	4 (11%)	3 (9%)	28 (80%)	0%	89%
8. Frequent Peace of Mind	4	1 (25%)	0%	3 (75%)	0%	75%
9. Nervous	28	3 (11%)	5 (18%)	18 (64%)	2 (7%)	82%
10. Fear	19	1 (5%)	5 (18%)	12 (63%)	1 (5%)	89%

11. Lack Attention	21	4 (19%)	2 (10%)	15 (71%)	0%	81%
12. Shyness	18	1 (5%)	5 (28%)	11 (61%)	1 (5%)	89%
13. Nightmares	12	0%	3 (25%)	9 (75)%	0%	100%
14. Forgetfulness	32	11 (34%)	1 (3%)	19 (59%)	1%	63%
15. Lack Confidence	25	3 (12%)	3 (12%)	18 (72%)	1%	84%
16. Inability to Concentrate	31	10 (32%)	7 (23%)	14 (45%)	0%	68%
17. Lost Memory	26	7 (27%)	1 (4%)	18 (69%)	0%	73%
18. Lack Interest	18	3 (17%)	3 (17%)	10 (55%)	2 (11%)	72%
19. Other	5	0%	1 (20%)	4 (80%)	0%	100%
Total Symptoms	409	61 (15%)	62 (15%)	275 (67%)	10 (2%)	82%

The health questionnaires definitely suggests a relationship between dental amalgams and health, especially mental health. Was it the mercury? Mental health symptoms were exhibited significantly more for subjects with amalgams compared to those without, 143 percent more (p=0.007). Eighty two percent of the subjects who had their amalgams removed found improvement in their mental health. It does appear that mercury in the brain is causing mental health symptoms in the early twenties. Is it accumulating in the brain over years and causing Alzheimer's disease at a later age?

CHAPTER FIVE

The Relationship Between Amalgam Mercury and the Heart, Eye, Hearing, Immunity Multiple Sclerosis, and Parkinson's Disease.

It May Surprise You

Before we discuss how dental amalgam is affecting the brain in the next chapter, it is important that we understand physiological and neurological changes that occur in people who have amalgams. We have explained how mercury vapor from dental amalgam is continuously escaping from the amalgam and entering the blood stream after inhalation. It can be transported to most parts of the body, especially the nervous system. Evidence suggests it is affecting the heart, blood pressure, the eye, hearing, immune system, and may be the cause of multiple sclerosis and Parkinson's disease.

THE CARDIOVASCULAR SYSTEM

Electrocardiography studies have shown that mercury affects the heart. One study examined rabbits who were affected by chronic metallic mercury vapor poisoning. The rabbits had ventricular depolarization with seven segments on the electrocardiogram being affected. In the 14 rabbits tested, they all had bradycardia, a slow heart rate.

Research on 42 mercury-poisoned patients in Iraq found they all had electrocardiogram changes. People exposed to low doses of mercury vapor can develop coronary insufficiency and pressure in the cardiac region, which are symptoms of micromercuialism.

Mercury is known to increase blood pressure. Eight mongrel dogs were injected subcutaneously with mercury chloride at a daily dose of 3mg/kg body weight for three days. The blood pressure increased between 10 to 70 mm/Hg in all dogs. A study of mercury exposed workers found 50 percent of the men and 68 percent of women had hypertension.

The rapid absorption of mercury into red blood cells results in inhibition of glucose and a loss of potassium which decreases osmotic fragility. Mercury can impair heme synthesis in erythrocytes (red blood cells) by inhibiting sulfhydryl enzymes. Individuals who worked daily with mercury found their percentage of hemoglobin decreased as their length of employment increased. With increasing length of employment, the number of erythrocytes decreased in most workers.

Three factors led us to study the relationship between dental amalgams and the cardiovascular system.

1. Mercury's affinity for the blood stream.

2. The relationship between mercury and the cardiovascular system.

3. The release of mercury vapor from the amalgam which enters the blood.

Studies have suggested many fillings could release up to 560 mg of mercury over several years. Mercury released from amalgam is in the form of elemental mercury vapor. An average of 75 to 80 percent of the mercury vapor inhaled is absorbed through the alveoli of the lungs, where it passes into the blood stream rapidly and completely. Immediately after exposure there is a much higher mercury content in the red blood cell than in the plasma. After oxidation, most of the mercury is bound to albumin or globulin in the blood. Once ionized, a mercuric ion can form reversible bonds with tissue ligands such as thiol groups in protein.

Methodology

An age and sex matched group with amalgams (n=50) was compared to a group without amalgams (n=51). The 20 males averaged 10.1 amalgams and the 30 females averaged 9.8 amalgams. The amalgam group's average age was 23.3 years and the non-amalgam group was 22.4 years.

Hair and urine were analyzed for mercury, and a blood profile was performed. Volunteers completed a health questionnaire regarding their cardiovascular history. Blood pressure and pulse were checked. Statistical analysis was done by the Colorado State University Statistics Lab.

Results

The levels of urine mercury were 210% higher in the amalgam group (p=0.0002) and hair levels were 26.5% higher (p=0.008). The urine mercury correlated with the number of amalgams (r=0.46, p=0.0001), as did hair mercury. The amalgam subjects had significantly higher blood pressure than the non-amalgam subjects. Both the systolic and diastolic pressures were higher in each sex of the amalgam group. The heart rate was significantly slower in the amalgam group

Blood Pressure (mm Hg)	Non-amalgam	Amalgam	P Value
Systolic (Both sexes)	100.71	106.41	0.0005
Diastolic	58.67	63.04	0.0015
Heart Rate	72.71	69.90	0.074

Hemoglobin levels were significantly lower in the amalgam group (p=0.016), and the red blood cell count was lower at the p=0.143 level. Total protein was significantly lower in the amalgam subjects (p=0.098), as was the hematocrit (p=0.013).

Significant correlations with urine mercury of the amalgam group was found with systolic blood pressure (negative correlation, p=0.023), with diastolic blood pressure (negative correlation, p=0.024), heart rate (positive correlation, p=0.024), with hemoglobin (negative correlation, p=0.003), with hematocrit (negative correlation, p=0.005), and red blood cells (negative correlation, p=0.001).

One cardiovascular health symptom, typical of mercury toxicity, was reported more frequently by the amalgam subjects, that being a greater incidence of heart or chest pains. The amalgam group was significantly more tired in the morning (p=0.001) and complained of tiring easily (p=0.007).

A non-controlled study on 86 subjects who had their amalgams removed said the symptoms of chest pain, tachycardia, heart murmur, and high and low blood pressure improved after amalgam removal.

Discussion

Approximately 90 percent of all people who have hypertension are said to have "essential hypertension," meaning hypertension of unknown origin. Perhaps mercury from dental amalgam should be considered a possible etiology. Of the seven subjects reporting high blood pressure before amalgam removal, five reported their high blood pressure improved or was eliminated after amalgam removal.

THE EYE

As mentioned earlier, when doing research for my master's degree in nutrition, I studied the relationship of nutrition and myopia (nearsighedness). I found that subjects without myopia had significantly higher hair mercury, a fact that I could not explain.

When doing my Ph.D. research on health and dental amalgams, the health questionnaire asked the subjects if they wore glasses for distance (myopia). A surprising answer occurred. There were 26 amalgam subjects who answered yes, 21 answered no, compared to 34 subjects without amalgams who answered yes, 12 answered no, which was a 31 percent difference at a p level of 0.083. Visual acuities in the amalgam group averaged 20/202 and in the non-amalgam group averaged 20/265 (p=0.148). Was mercury from amalgams preventing myopia? This would explain my hair mercury finding.

The study for my master's degree involved 25 subjects with myopia, including 12 males and 13 females. They were compared to 25 non-myopes consisting of 11 males and 14 females matched for age. The non-myopia hair levels were 0.141 ppm, compared to myopes 0.101 ppm (p=0.005), a 40% difference. Both female non-myopes and male non-myopes had higher levels of mercury.

A summary of two studies, without amalgams (n=124) and with amalgams (n=123) found the without-amalgam group reported 34 percent more frequency wearing glasses for distance (p=0.007).

Two studies, one on rats and one on monkeys, found that mercury does have an affinity for the eye. The distribution of inhaled radioactive metallic mercury vapor was found to be higher in the retina, pigment epithelium, and the choroid. The eye is considered to be part of the brain.

During the 1950's, Dr. C. Desusclade, a French physician used vitamin E and mercury to retard the progression of myopia. Rarely did myopia progress for children who followed his treatment.

One cause of myopia is a disassociation of scleral collagen fiber cross-linking, which weakens the eyeball and results in elongation of the eye causing myopia. If mercury could increase the collagen of fibrils, perhaps it could retard the progression of myopia. A study using rabbits found that mercuric chloride administered to the animals increased the amount of collagen fibrils in the basement membrane of the kidney, as well as the ileum and colon. In that study, collagen in the eye was not tested.

In summary, it does appear that dental amalgam mercury slows the progression of myopia.

HEARING

One of the many symptoms of mercury toxicity is hearing loss. As will be discussed in the next section, there is evidence that amalgam mercury is associated with multiple sclerosis (MS). The toxic effect of mercury can demyelinate nerve fibers and slow the nerve conduction velocity. Epidemiological studies have correlated dental fillings to MS. Our research involved testing for hearing of MS subjects. Seven female MS subjects were tested for hearing at threshold frequencies of 250, 500, 1,000, 4,000, and 8,000 Hz before amalgam removal. Six of the seven subjects showed improvement in the right ear and five of the seven showed improvement in the left ear after amalgam removal. The total frequencies were averaged before amalgam removal and compared to after amalgam removal. Hearing improved an average of 8 db (p=0.02).

One of the characteristics of nerve deafness is a decline for all frequencies. There was significant improvement in all frequencies in the right and left ear after amalgam removal.

The results suggested that the hearing loss was due to nerve damage. Amalgam Mercury can lead to nerve damage by affecting RNA protein synthesis, by reducing nerve conduction velocity, by demyelination of the nerve fiber, by increasing the threshold for excitation, by blocking action potential, and by affecting the neurotransmitter secretion or the receptor site. The evidence suggests that mercury from dental fillings can reduce hearing.

MULTIPLE SCLEROSIS

Epidemiological studies have linked dental caries with multiple sclerosis. In Australia, the rate of death due to MS is linearly related to the number of decayed, missing, and filled teeth (p=0.001). The highest incidence of MS is found in Northern Ireland and the Scottish islands of Orkney and Shetland. These areas of the world are ranked highest in dental caries. In 1966, Dr. Bausch, a Swiss neurologist recognized the possibility of dental amalgam fillings being a source of mercury that may cause MS. He believed MS was a neuro-allergic reaction to mercury from dental amalgams.

Dental caries increase with geographical latitude leading researchers to theorize that the incidence of dental caries increases with decreasing sunlight, with sunlight providing protection against dental caries. The sunlight hypothesis has been applied to MS as well. Sunshine provides vitamin D, and it is known that vitamin D is anti-inflammatory. It is also known that mercury causes inflammation. In 1983, Dr. Ingall theorized that the correlation between dental caries and MS might be related to dental amalgam mercury.

Multiple sclerosis was first reported in the literature during the 19th century shortly after the dental amalgam was introduced to the public as a restoration for dental caries. In 1833, the amalgam was introduced to the United States by the Crowcour brothers who came to New York from Great Britain. The lesions of MS were probably first described as early as 1838. When amalgam mercury started being widely used, MS also became more prevalent.

It is known that T-lymphocytes, which fight infections, are suppressed in MS patients with the subclass CD8 (T-8 suppressor lymphocytes) being reduced during exacerbation of MS symptoms. According to Eggleston, the T-lymphocyte count fell when amalgams were placed in the mouth. My study at Colorado State University found that 20 females without amalgams had a significantly higher percentage of T-lymphocytes when compared to a control group of 20 age-matched females with amalgams.

An elevated IgG (immunoglobulin G, an antibody) in the cerebral spinal fluid is a diagnostic test for MS. It has been shown that when T-8 suppressor lymphocytes are reduced, the levels of IgG are increased. A study at Colorado State University by the author compared 14 females with amalgams to 14 non-amalgam females. There were significant correlations with IgG, IgA, and IgE to the number of amalgams and urine mercury.

Studies have shown that the blood-brain barrier (BBB) is impaired in MS patients, and it is thought this may be a contributing factor in MS. Chang found that when mercury ions are absorbed into the blood stream in minute amounts, they are capable of impairing the blood-brain barrier in rats. The study found that mercury enters the parachyma of the central nervous system. When mercury is bound to the blood-brain barrier, it causes a leaky membrane that allows viruses and foreign protein to pass through. Alterations to the BBB is thought to be of crucial importance in MS because this could be the step which permits entry of immuno-competent cells, antibodies, or other effects which lead to myelin destruction.

Chang demonstrated that after HgCl2 poisoning, a large axonal space was created in many axons as a result of detachment of the axonal shrinkage, and there was occasional destruction. Degeneration was observed along with axonal collapse and myelin destruction. Demyelination of the nerve is the pathogenesis of MS.

One of the diagnostic findings of MS is a reduction of nerve conduction velocity as measured by the visual evoked response (VER) test. One characteristic of mercury toxicity is a slowing of the nerve conduction velocity.

A study by Knolle and Gunther reported that not all MS patients had amalgam fillings when the disease began. Their study found that out of 100 MS patients, 11 had been previously treated with mercury ointments, so other mercury sources must be considered. Currier noted the unusual amount of dental work among MS patients, but he also noted a lack of fillings in some patients.

Selenium is one of nature's protectors against the harmful effects of mercury. Schlin and Irvine provided evidence of an association between geographic areas with low selenium and a high incidence of MS. Selenium is the main functioning element in glutathione peroxidase, which destroys free oxy-radicals before they can attack the cellular membrane. If deficient, lipid peroxidation of nerve fibers would result. My research at CSU found mercury produces free oxy-radicals. Other studies have confirmed this.

Aholrot-Westerlund found that mercury levels in the cerebral spinal fluid were eight times higher in MS patients compared to a control group without MS, suggesting that mercury was associated with MS.

Many MS patients in the 1980's had their dental amalgams removed after hearing about the connection with amalgam mercury. Our study was conducted on 50 MS subjects who had their amalgams removed and were compared to 47 MS subjects with amalgams. For the blood study, the average age of the MS non-amalgam group was 41.8 years and the average age of the MS amalgam group was 42.9 years. Their amalgams were removed 2.75 years before the study. Both groups consisted of 20 females and 4 males in the blood study.

Blood Results

1. The MS group with amalgams had significantly higher blood urea nitrogen compared to the control group with amalgams removed.

2. The electrolyte potassium was significantly higher in the amalgam removal group.

3. The calcium level was significantly higher in the amalgam removal group.

4. The MS subjects without amalgams had a significantly higher red blood cell count, hematocrit, and hemoglobin levels than those with amalgams.

5. The thyroid gland also appeared to be affected as thyroxine (T-4) levels were significantly lower in the MS group with amalgams.

6. The total T-lymphocytes were significantly higher in the amalgam removal group.

7. The T-lymphocyte differences were more pronounced in the females. The T-8 suppressor and T-8 suppressor percentages were significantly higher in the female MS amalgam removal group, while the T-4 helper/T-8 suppressor ratio was significantly lower. The subclass of T-8 suppressor

lymphocytes is suppressed in MS subjects especially when there is an exacerbation of MS symptoms. One theory of the etiology of MS is that a suppression of the T-8 lymphocytes results in an overactive immune system that attacks the nervous tissue resulting in sclerotic lesions. T-8 suppressor cells also control the antibodies, such as IgG. A reduction of T-8 suppressor cells could account for the elevated IgG levels found in the cerebral spinal fluid of MS subjects. In a non-MS study, we found a significant correlation between urine mercury and IgG levels (p=0.048).

8. The MS male removal group had significantly higher T-lymphocytes compared to the male MS amalgam group.

9. MS subjects had significantly higher levels of hair mercury compared to a control group of age and sex matched non-MS subjects.

10. Fourteen of 45 MS subjects (31%) felt their symptoms of MS got better following amalgam removal while three subjects (7%) felt it was eliminated. There were 15 (33%) who felt there was no change, and 29% felt the MS got worse.

11. When comparing present neuromuscular symptoms, the MS group with amalgams had 17% more symptoms per subject (p=0.097). The symptoms of paresis of arms, clumsy fingers, arm tremors, leg tremor, and hand tremor were reported significantly more in the group with amalgams.

12. The comparison between symptoms during the past 12 months found the MS group with amalgams having 33.7% more symptoms per subjects (p=0.0037) compared to the amalgam removal group.

The symptoms of muscle pain, unsteady walking, lack of balance, arm weakness, paresis of arm, spastic arm, tingling arm, numb arm, clumsy fingers, arm tremors, leg tremors, and hand tremor were significantly higher in the MS amalgam group.

13. MS subjects with amalgam removal reported 37% of the symptom of frequent night urination improved or was eliminated compared to those with amalgams. They reported 46% less frequent day urination.

14. The amalgam removal group reported 39% of their chronic fatigue symptoms had improved or were eliminated, and 35% felt they did not tire as easily.

15. The study found that the MS group with amalgams had 33 percent more exacerbation of neuro-muscular symptoms compared to the MS removal group within the past 12 months. This suggested that mercury could be an initiator of the exacerbations. The mercury released from the amalgams may initiate an autoimmune response, and exacerbate the symptoms.

16. The significant higher blood urea nitrogen in the MS amalgam group suggested that mercury may be affecting the kidney. An increased BUN is a finding found in chronic renal disease. Studies have found that mercury has a strong affinity for the kidney and accumulates there with age. This study showed that the amalgam removal group had a decrease in symptoms of frequent night urination by 37% and a decrease in frequent day urination of 46% following amalgam removal.

The considerable number of significant findings suggest there may be a relationship between amalgam mercury and MS.

Nerve Conduction Velocity

One of the hallmark traits of multiple sclerosis is reduced nerve conduction velocity as measured by the visual evoked response. A characteristic of mercury toxicity is that it slows nerve conduction velocity.

The next phase of the MS study examined nerve conduction velocity before amalgam removal and after removal. The study involved seven female MS subjects averaging 37.7 years in age. An MRI test was performed on all subjects which confirmed central nervous lesions for all patients in the study. Before removal of amalgams, the seven MS subjects underwent a visual evoked response (VER) test to measure nerve conduction velocity. Each subject was tested with two trials in each eye. Following the VER, each volunteer was assigned to a mercury-free dentist in Colorado who removed their mercury amalgams. Six months after amalgam removal, the VER test was repeated.

Of the seven subjects, five had shown a decrease in the mean latency of the P1 and N2 components of the VER following amalgam removal. The average change for P1 was 23.1 milliseconds (p=0.011), and for N2, it was 31.7 milliseconds (p=0.026). In other words, the mean latencies decreased significantly in both eyes, meaning the nerve conduction velocity significantly increased after amalgam removal.

Thyroid

Other characteristic symptoms of MS are cold hands and feet. Sixty-seven percent of the MS subjects in the study reported these symptoms. A lower body temperature often suggests there may be a thyroid gland disorder. Mercury is known to affect the thyroid gland by inhibiting thyroxine. A common symptom of hypothyroidism is a low body temperature. The MS amalgam group had significantly lower thyroxine levels than the amalgam removal group (p=0.012).

The author recently had a scientific paper published entitled *A Hypothesis and Additional Evidence That Mercury May Be An Etiological Factor In Multiple Sclerosis*. This paper identified factors identified with MS as listed in *The Encyclopedia Of Multiple Sclerosis*. These factors were cross referenced with mercury on

PubMed to identify scientific articles that determined if mercury could explain these changes. All 32 physical factors associated with MS could be explained by mercury toxicity. All 7 mental health factors associated with MS could be explained by mercury toxicity as well as all 11 ocular symptoms. The paper confirmed the earlier evidence that mercury may be associated with MS.

PARKINSON'S DISEASE

A similar research protocol was conducted with Parkinson's disease (PD). Mercury was cross-referenced with PD regarding 93 factors associated with PD found in *The Encyclopedia of Parkinson's Disease.*

The cause of PD is a reduction in the neurotransmitter dopamine. In PD, the cells in the substantia nigra reduce production of dopamine. The primary cause is the death of dopamine neurons in the brain.

Mercury is known to decrease dopamine levels and cause dopamine neuron degeneration. The cardinal symptoms of PD include resting tremor, bradykinesia (slowness of movement), muscle rigidity, and postural instability. All four cardinal symptoms of PD can be caused by mercury toxicity. Of the 93 factors associated with PD, we found that 80 have been shown that mercury could cause these symptoms. The remaining 13 factors we could not find anything in the literature that associated them with mercury, one way or another.

Dr. Hal Huggins (chapter 9), a mercury-free dentist who practiced in Colorado Springs, CO, wrote in his book that several of his PD patients improved significantly following amalgam removal My scientific paper on PD is currently under review at the time of writing.

Dr. Joachim Mutter (Chapter 8), an environmental physician in Germany, reports of his Parkinson's disease patients recovering after amalgam removal and detoxifying with DMPS, DMSA, and Imininix, plus good nutrition.

Conclusion

Evidence has been presented how dental amalgam mercury affects the heart, the eye, hearing, immunity, thyroid and may be the the cause of multiple sclerosis and Parkinson's disease. The eye is considered part of the brain, and evidence suggests that mercury is affecting the eye.

Our studies have suggested mental health symptoms are the most frequent symptom of amalgam mercury toxicity, as will be discussed in the following chapter. This means mercury in the brain originates from dental amalgam, and we then ask, could dental amalgam mercury be a cause of Alzheimer's disease?

CHAPTER SIX

The Effect of Amalgam Mercury on Mental Health

Mercury's Affinity for The Brain

On my journey to discovering the health effects of amalgam mercury, the most consistent finding was that it seemed to affect mental health the most. The findings strongly suggest that amalgam mercury is entering the brain, and causing a variety of mental health problems. Significant findings were found with depression, anger, irritability, schizophrenia, and bipolar disorders. Remember, that once inside the brain, mercury can be entrapped for several decades or longer. The following is a summary of the various studies we conducted using standardized mental health testing under the guidance of John Motl, M.D., a Fort Collins psychiatrist. All the studies were published in scientific journals.

DEPRESSION AND EXCESSIVE ANGER

This study involved 25 women (mean age 35.3 years) with dental amalgams compared to 23 women (mean age 39.6 years) without amalgams. The amalgam group averaged 13 amalgams. The subjects were given standardized mental health questionnaires, which included the Beck Depression Inventory and the State Trait Anger Expression Inventory.

Depression

The Beck Depression Inventory consists of 21 questions. The mean total score for the group with amalgams was significantly higher than the mean for subjects without amalgams (p=0.04). The amalgam group had significantly higher scores on 9 of the 21 questions and scored higher on the remaining questions but not at the significant level, i.e. p>0.10. The nine symptoms which rated significantly higher were pessimism (p=0.05), guilt (p=0.002), punishment (0.07), self-accusation (p=0.10), body-image (p=0.08), work difficulty (p=0.09), insomnia (p=0.04), fatigability (p=0.04), and somatic preoccupation (p=0.03).

Anger

The State-Trait Anger Expression Inventory found that the amalgam women group had significantly higher mean scores of angry temperament and a propensity to express anger without provocation. Their scores were also higher on "anger out" meaning how often they expressed anger, and also higher in "anger expression," meaning the index of frequency that the anger is expressed.

The amalgam group reported significantly higher scores on being quick tempered (p=0.06), expressing anger aggressively (p=0.02), experiencing intense angry feelings (p=0.08), slamming doors when angry (p=0.03), saying nasty things (p=0.07), losing their temper (p=0.06), being hot headed (p=0.06), and telling someone they are annoying (p=0.003). The non-amalgam group had significantly higher scores in being patient with others (p=0.10), calming down faster than most people (p=0.06), and being furious when criticized (p=0.03).

Oral Cavity Mercury

This study also compared mercury vapor levels in the oral cavity. Volunteers in the amalgam group exhibited a mean mercury vapor of 0.02 mcg/m3 in the oral cavity before chewing. After chewing, the mean vapor increased to 0.07 mcg/m3, a 350 % increase. The subjects without amalgams had a pre-chewing level of 0.01 mcg/ m3 and a post chewing level of 0.017 mcg/m3, a 58.8 % increase. The difference between the two groups was significant at the p=0.001 level.

A SECOND STUDY COMPARING MENTAL HEALTH OF SUBJECTS WITH AND WITHOUT AMALGAMS

The second study compared the mental health status of 50 subjects with amalgams (30 females and 20 males) to 51 subjects with no amalgams (30 females and 21 males). The average age of the amalgam group was 23.28 and the non-amalgam group was 22.35 years. The males averaged 10.1 amalgams and the females 9.8 amalgams.

A hair sample from the nape of the neck and urine specimen was analyzed for mercury. Two mental health questionnaires were given to these subjects. A mental health questionnaire was also sent to 86 subjects who had their amalgams removed that included 60 females and 26 males who averaged 40.41 years in age.

Mercury in Tissue

	Amalgam	Non-amalgam	% Different	P value
Hair Mercury	1.47 ppm	1.13 ppm	26.5%	0.008
Urine Mercury	3.7 ppb	1.23 ppb	300 %	0.0002

Health Questionnaire

The non-amalgam group rated themselves to be significantly happier, 8.48 to 8.02 on a scale of 1 to 10,(p=0.047), greater peace of mind 8.02 to 7.54 (p=0.075), good reading comprehension 41 non-amalgam subjects to 35 amalgam subjects (p=0.04). The amalgam group reported significantly more symptoms of sudden anger (1 non-amalgam subject to 10 amalgam subjects, p=0.0046), more depression (7 non-amalgam subjects to 18 amalgam subjects, p=0.008), and more irritability (9 non-amalgam subjects to 19 amalgam subjects, p=0.018). Of the seven questions in the emotional and mental questionnaire, the amalgam group reported 145 % more symptoms (p=0.007).

In the health questionnaire sent to the 86 subjects who had their amalgams removed, there was a great improvement in mental health. It is important to note that this aspect of the study was not controlled, and there may have been a placebo effect.

Not all 86 volunteers responded to each question. 67 of 84 (80%) reported they felt better after amalgam removal and 91% were glad they had their amalgams removed. After removal, 68% said they were more tolerant of stress. 63.8% of the subjects reported a health improvement, and no one reported a deterioration of health. They reported their health improved by 26.2%. Happiness and peace improved in 58% of the subjects, being 26.4% more happy.

The 86 subjects reported a total of 409 psychological symptoms. Within an average of 10 months following amalgam removal, 67% of their symptoms improved, 15% were eliminated, 15% did not change, and 3% got worse.

Of 31 subjects reporting sudden anger, 90% stated that the anger emotion had improved or was eliminated. Depression was reported by 38 subjects before removal, and 84% of depression symptoms were improved or eliminated. Irritability, a common symptom of mercury exposure, improved in 83% of 35 subjects after amalgam removal. Frequent anxiety improved or was eliminated in 89% of 35 subjects, while nervousness declined in 82% of 28 subject. Nightmares were reduced in all 12 subjects who had them. In all 10 subjects who reported suicidal tendencies, 100% said the condition was improved or eliminated. Memory improved in 19 (73%) of 26 subjects who reported poor memory. Confidence increased in 21 (84%) of 25 volunteers who had their amalgams removed. Fatigue symptoms improved or were eliminated in 53% of 34 subjects who reported fatigue, and they were not as tired in the morning.

MULTIPLE SCLEROSIS AND MENTAL HEALTH

In the last chapter we discussed the relationship between mercury and MS. Besides noting the effects on MS subjects physical health, we studied its effects on mental health.

Multiple sclerosis is associated with a deterioration of mental health. Researchers are unsure whether this worsening of mental health results from stress of having a debilitating disease such as MS or is it part of the disease itself. Evidence now suggests the accompanying psychological effects may be physiological in origin.

In a study in 1969, 108 MS subjects were compared to 39 subjects with muscular dystrophy. Depression was manifested in 27% of MS patients compared to 13% of the muscular dystrophy subjects. In about 5% of MS subjects, depression may be the first symptom. Often cognitive decline accompanies MS, ranging from 67% to 28% who suffer from it. Patients with long established MS have difficulty in learning and recalling verbal and nonverbal material.

Emotional distress is strongly associated with the exacerbation of MS. It is uncommon when MS is in remission. The major emotional disturbances accompanying MS include depression, negative self esteem, and psychosis.

Twenty-four dentists in the state of Colorado, who stopped using amalgams, were asked to supply names of subjects with MS who had their amalgams removed. From these names, we recruited 50 volunteers. Newspaper ads recruited 47 MS subjects to serve as a control.

Three standardized psychological tests were given the subjects including the Beck Depression Inventory, the State Trait Anger Expression Inventory, and the Symptom Check List 90-Revised. Statistical analysis was done by the Statistics Lab at Colorado State University.

A mental health questionnaire containing 18 symptoms was given to each group asking if the subjects experienced these symptoms in the last 12 months. The MS subjects with amalgams reported 43.5% (p=0.005) more symptoms per subject than the group with amalgams removed. The symptoms of sudden anger, depression, "wish I was dead," irritability, trouble with decisions, forgetfulness, inability to concentrate, and memory loss were reported significantly more in the MS group with amalgams.

A third mental health questionnaire involved 11 symptoms typically found in MS patients. The amalgam group had 36.5% significantly more symptoms compared to the control, including symptoms of depression, mental confusion, irritability, euphoria, and stupor.

A second mental health questionnaire involving 17 symptoms was given to the MS subjects who had their amalgams removed. Of the 289 symptoms reported before amalgam removal, 42% were improved, 12% were eliminated, 26% did not change, and 19% got worse. Sudden anger was reported as either being eliminated or improved in 75% of the subjects. Depression was improved in 48% of the subjects and eliminated in 17%. Suicidal tendencies were either eliminated or improved in 7 of 7 subjects who reported the symptom. Anxiety symptoms were improved in 48% and eliminated in 14%. Irritability symptoms were improved or eliminated in 67%. This questionnaire was not given in a controlled setting but in a before and after manner. It was an average of 2.8 years after amalgam removal.

Beck Depression Inventory

Amalgam MS subjects tested for significantly more symptoms of depression (p=0.07) than those without amalgams. Of the 21 categories of Beck scores, the MS amalgam group rated higher in 15. The categories of sadness, dissatisfaction, self-dislike, suicidal ideas, and indecisiveness were reported significantly more in the MS amalgam group.

State Trait Anger Expression Inventory

On the State-Trait Anger Expression Inventory, 3 of 8 anger categories were significantly higher in the MS amalgam group. They included trait anger, anger temperament, and anger reaction. Four others rated higher, but not at the significant level.

Symptom Check List 90-Revised (SCL-90)

The MS amalgam removal group had significantly better overall scores in the SCL-90 test (p=0.02). Means for the categories of obsession-compulsion, depression, anxiety, hostility, and psychotism were significantly higher in the MS group with amalgams. Those with amalgams also had significantly more symptoms of poor appetite, thoughts of death, awakening in early morning, and disturbed sleep.

Depression is one of the most common symptoms of both MS and mercury toxicity. This was confirmed on the Beck Depression Inventory, Symptom Check List - 90, and general questionnaires, suggesting mercury may be an initiating factor.

Anger is another characteristic symptoms of mercury toxicity which gave rise to the expression "mad hatter." Earlier studies had indicated that subjects with amalgams reported significantly more symptoms of sudden anger (p=0.005). The MS subjects also reported more symptoms of anger, and the SCL-90 test showed they had higher scores of hostility. Trait anger scores were significantly higher in the amalgam group.

Amalgam MS subjects reported poorer cognitive abilities, such as poor short term memory. After amalgam removal, 36 percent said they were not as forgetful, 35 percent reported an improvement with memory, and 50 percent were able to concentrate better.

The studies strongly suggested that poor mental health is associated with mercury from dental amalgams. The next step in our research was to study manic depression disorders and schizophrenia.

BIPOLAR DISORDER (MANIC DEPRESSION)

If mercury was entrapped in the brain, could it be causing other mental disorders? Depression is one of the most consistent symptoms of mercury toxicity. It has been postulated that depression from amalgam mercury may be the result of affected neurotransmitters in the brain, as mercury has been shown to affect the same neurotransmitters whose disruptions causes manic depression. This study hypothesized that if dental amalgams containing mercury were removed, depression and mania might subside.

The study compared 11 bipolar subjects with amalgams (5 females and 6 males) with a mean age of 40.8 years to a bipolar control group (4 females and 5 males) with a mean age of 36.4 years. The control bipolar group also had amalgams but was told that a sealant was going to be placed on the amalgam to prevent mercury from escaping. This served as the placebo, but there is no sealant to prevent mercury from escaping.

Methodology

The subjects completed three psychological tests which included the Minnesota Multiphasic Personality Inventory-II (MMPI-II), the Millon Clinical Multiaxial Inventory-II (MCMI-II), and the Symptom Check List -90 (SCL-90) before and after treatment. Mercury vapor in the oral cavity was measured by a trained professional with a Jerome Mercury Vapor Analyzer before chewing a piece of gum and again ten minutes after chewing.

Each of the 11 subjects in the treatment group had their amalgams removed by a different mercury-free dentist in Colorado. The nine control subjects were given a placebo sealant in a dentist's office. These subjects were told the sealant would prohibit mercury from escaping. The amalgam-removal group averaged 10.4 amalgams and the placebo/sealant group averaged 9.6 amalgams. A health questionnaire was given before and after treatment.

The amalgam-removal group was given supplemental nutrients, which included a multiple vitamin, vitamin C, zinc, vitamin E, garlic, and glutathione peroxidase, supplements that help detoxify mercury by helping the body neutralize mercury. The control group was given one multiple vitamin and a mineral tablet.

Statistical analysis was conducted by the Statistics Laboratory at Colorado State University.

Results

Minnesota Multiphasic Personality Inventory II (MMPI II)

The MMPI consists of 567 questions making up a total of 87 scales. The amalgam removal group showed a statistically significant improvement in 47 of the scales when compared to the amalgam-sealant/placebo control group. An improvement was found in another 20 scales over the control group but not at a significant level. There was a statistical improvement in six scales in the control group.

Millon Clinical Mutiaxial Inventory

The MCMI II consists of 175 true and false questions that make up 20 scales. The amalgam removal group showed a statistically significant improvement in 4 scales compared to the amalgam-sealant/control group. There was a significant improvement in the Clinical Personality Pattern category and the Severe Personality Pathology.

Symptom Check List - 90

The SCL 90 test consists of 90 questions that are utilized to evaluate nine various dimension. The amalgam removal group showed a significant improvement in all nine dimensions when compared to the control. The nine dimensions included somatization (p=0.007), obsessive-compulsive tendencies (p=0.008), interpersonal sensitivity (p=0.049), depression (p=0.007), anxiety (p=0.097), hostility (p=0.01), phobic anxiety (p=0.034), paranoia (p=0.061), and psychotism (p=0.011). The total improvement was significant at the p=0.008 level.

Oral Cavuty Mercury

Oral cavity mercury vapor was measured before treatment in both groups combined (n=19). It increased 290 percent after chewing for 10 minutes from 0.024 mcg/m3 to 0.70 mcg/m3. Remember, the control group still had their amalgams.

Health Questionnaire

The amalgam removal group averaged 36.5 symptoms/subject before treatment and 21.2 symptoms/subject after amalgam removal, a 41.9 percent decrease. The placebo/sealant group averaged 18.8 symptoms/subject before treatment and 20..2 symptoms after treatment, a 7.4 percent increase.

Discussion

Most psychotherapists believe that bipolar disorder is a result of a neurochemical imbalance of the brain, affecting neurotransmitters. The catecholamine hypothesis states that depression is associated with a deficiency of catecholamines, particularly norepinephrine at important receptor sites in the brain. Manic elation, conversely may be associated with a excess of such amines.

In 1975, it was proposed that dopamine dysfunction might be a component of depression and mania as well. It was hypothesized that mania is associated with increased dopaminergic neuronal activity while depression was associated with decreased dopaminergic activity.

The literature suggests that bipolar disorders may result from disruption of the secretion and reception of norepinephrine, acetylcholine, serotonin, and dopamine. If one could determine what causes this dysfunction, it might explain the etiology of manic depression. Research has shown that mercury affects norepinephrine, acetylcholine, serotonin and dopamine. Mercury has the ability to affect the secretion of these neurotransmitters and their uptake at the receptors. Perhaps this might explain why the mental health status of bipolar subjects who had their amalgams removed improved significantly.

SCHIZOPHRENIA

Schizophrenia is another mystery of mental health. What causes schizophrenia? Evidence suggests a dysfunction with the neurotransmitters such as dopamine, monoamine oxidase, and serotonin, which have all been implicated, and all are affected by mercury.

Methodology

In this study eight schizophrenia subjects participated in the study, including five males and three females, with an average age of 40.86 years. They averaged 23.5 amalgam surfaces.

The subjects were given three psychological tests, the Minnesota Multiphasic Personality Inventory (MMPI-II), the Millon Clinical Multiaxial Inventory (MCMI -2), and the Symptom Check List- 90 (SCL-90). They also completed a health questionnaire, and blood tests were administered. Each volunteer had all their dental amalgams removed by a Colorado dentist, and each received nutrition supplements to neutralize the effects of mercury. Psychological testing and blood tests were repeated six to nine months following amalgam removal. Statistics were completed by the Statistics Laboratory at Colorado State University.

Results

Minnesota Multiphasic Personality Inventory II (MMPI II)

The MMPI-2 consists of 567 questions that make up a total of 61 component scales and 20 subscales. Significant improvement at the p=0.10 level was found in 41 of the 61 components scales. Improvement at the p=0.05 level was found in 33 of the 61 component scales. Twelve of the 20 sub scales improved significantly at the p=0.05 level.

The sub scales that improved significantly were schizophrenia, hysteria, psychopathic deviate, paranoia, social introversion, fears, health concerns, bizarre mentation, anger, cynicism, low self-esteem, social discomfort, and family problems.

Millon Clinical Multiaxial Inventory -2

The MCMI-2 consists of 175 true and false questions that made up 25 diagnostic scales and five categories. Sixteen of the 25 diagnostic scales improved at the p=0.05 level and 17 improved at the p=0.10 level.

The scales that improved were disclosure, debasement, schizoid, avoidant, dependent histronic, narcissistic, antisocial, aggressive-sadistic, compulsive, passive-aggressive, self-defeating, anxiety disorders, bipolar, alcohol dependent, drug dependent, and thought disorder.

Symptom Check List-90

The SCL-90 is composed of 90 questions comprising nine dimensions. Improvement in 4 of 9 dimensions was significant at the p=0.10 level and one of nine at the p=0.05 level. Improvement was found in dimension categories of obsessive compulsive, interpersonal sensitivity, depression, and psychotism.

The health history questionnaire consisted of 177 questions and 13 categories. The mental health category improved by 43 percent and the dental symptom category improved by 46 percent. A nine percent improvement in somatic health symptoms was found. No significant difference was found between before and after blood tests.

Discussion

A significant improvement did occur in schizophrenia subjects after amalgam removal, demonstrated by improvement in all three psychological tests. The study suggested mercury may play an etiological role in schizophrenia. Psychotherapists believe schizophrenia is one of the most crippling psychological disorders, with its characteristic psychotic symptoms that result in disturbance of thought process and perception. As a result, social and occupational functioning severely deteriorates. Schizophrenia usually appears during adolescence or early adulthood up to middle age. Symptoms may include delusions, thought disturbance, hallucinations, and difficulty with goal directed activity. Paranoid schizophrenia involves preoccupation with one or more delusions or auditory hallucinations related to a single theme. Associated with schizophrenia are anxiety, anger, or possible violence. The exact cause of schizophrenia is not known. In several studies there was an inverse correlation between degree of monoamine oxidase (MAO) and degree of symptomology. It has been demonstrated that methyl mercury causes dose-related decreases in monoamine oxidase. Animal studies also show that mercury inhibits MAO.

The dopamine hypothesis postulates that schizophrenia is caused by excess dopamine dependent neuronal activation in the brain. Postmortem studies have found significant elevations of dopaminergic receptor sites in schizophrenia, especially DA-2 receptor sites. It is known that mercury does affect dopamine and its receptor sites.

Schizophrenia And Mercury

We conducted another study on schizophrenia, similar to the one on Alzheimer's disease that you will read about in the next chapter. All the symptoms and physiology of schizophrenia were identified in the book entitled *The Encyclopedia of Schizophrenia and Other Disorders* written by Richard Noll, Ph.D. All were cross referenced with mercury articles found in PubMed, the leading summary of scientific articles. Nearly all the changes that occur in schizophrenia could be explained by mercury toxicity.

There are four hallmark symptoms of schizophrenia called the "4 A's,"which include affective disturbance, ambivalence, association disturbances, and autism. All could be explained by mercury toxicity. There were 39 mental health symptoms associated with schizophrenia including hallucinations and delirium that could be explained by mercury.

Fifteen epidemiology findings were identified in the book and most could be explained by mercury toxicity. Eight theories on the cause of schizophrenia were identified and all could be explained by mercury toxicity. It does appear that schizophrenia could be caused by mercury toxicity.

This chapter has demonstrated the strong influence that amalgam mercury has on the brain by causing many mental issues. Bipolar disorder, depression, anger, irritability, and schizophrenia are all possible disorders that may relate to mercury. With mercury accumulating in the brain throughout life, we again ask, could it also be causing Alzheimer's disease?

CHAPTER SEVEN

Evidence That Dental Mercury Causes Alzheimer's Disease

The Scientific Proof

For several years I was gathering information for a book I was writing on ways to prevent Alzheimer's disease (AD). I checked out a book from the Colorado State University library entitled *Encyclopedia of Alzheimer's Disease* written by Elaine Moore. Elaine had done an excellent job on researching the pathology and physiology of AD for her book. She had compiled 70 changes that occurred with AD. Several decades earlier I had received a grant looking at the relationship between dental amalgam mercury and Alzheimer's disease but was unable to recruit AD subjects. I thought it would be interesting to see if these changes could be explained by mercury. I then crossed-referenced these changes on the internet with mercury, mainly through PubMed, the primary source of scientific research articles. To my astonishment, every one of these AD changes could be explained by mercury toxicity. The Colorado State University Statistics Laboratory did statistical analysis on this, and the probability of this happening by chance was less than 1 in 10,000 (P=0.0001). It was at that moment that

I realized I could prove mercury caused AD.

I didn't know what I should do next. So I decided to write a letter to Paul Allen, the co-founder of Microsoft, to see if he would be willing to fund research regarding the relationship between amalgam mercury and AD. His mother had died from AD, so Paul had founded a brain research institute to help find the cause. I did not hear back from Paul. He passed away in 2018 from cancer.

I then decided to write a scientific paper on the evidence that I found regarding mercury and AD. In my literature search I came across an environmental physician who had been doing research with amalgam mercury toxicity and health. His name was Joachim Mutter from Germany. He also believed there was a connection between AD and mercury from amalgams. Dr. Mutter had a scientific paper published describing how an ALS (Lou Gehrig's Disease) patient recovered following amalgam removal. He also had experience of removing amalgams from an AD patient. Dr. Mutter agreed to be a co-author and suggested that we also ask Dr. Johannes Naumann and Dr. Harald Walach to be co-authors, as they were two academics who were involved in mercury toxicity research. They both agreed to co-author. I also asked Elaine Moore to be a co-author, as she had written the *Encyclopedia of Alzheimer's Disease* book. I discovered she was also from Colorado.

Dr. Mutter warned me that if I mentioned dental-amalgam mercury toxicity in the paper, scientific journals would not touch it. I told Dr. Mutter I would like to try and mention dental amalgams, and see if we could get it published with all the evidence. Dr. Mutter was correct. A number of top journals turned it down. I then condensed the paper and removed all but several references to dental amalgams.

The scientific paper was published in December of 2019 by the *International Journal of Environmental Research and Public Health*, a very respected peer reviewed journal. The title of the paper was *A Hypothesis and Evidence that Mercury May be an Etiological Factor in Alzheimer's Disease* which follows.

A HYPOTHESIS AND EVIDENCE THAT MERCURY MAY BE AN ETIOLOGICAL FACTOR IN ALZHEIMER'S DISEASE

Robert Siblerud, Joachim Mutter, Elaine Moore, Johannes Naumann, and Harald Walach

Abstract

This study explored the hypothesis that mercury may be a cause of Alzheimer's disease by cross referencing the effects of mercury with 70 factors associated with AD. The study found that all these factors could be attributed to mercury toxicity. Hallmark changes in AD include plaques, beta amyloid protein, neurofibrillary tangles, phosphorylated tau protein, and memory loss, all explained by mercury.

Acetylcholine, serotonin, dopamine, glutamate, and norepinephrine are inhibited in both AD and mercury toxicity. Similar enzyme dysfunction occur in AD and mercury toxicity, involving gamma secretase, nitric oxide synthetase and acetyl choline transferase.

Immune responses seen in AD also occur in mercury toxicity, including complement activation and cytokine expression. Genetic factors in AD are also associated with mercury. Apolipoprotein E 4 allele increases mercury toxicity. Mercury can inhibit DNA synthesis in the hippocampus, and is associated with AD genetic mutations of presenilin 1 and 2.

Abnormalities of minerals and vitamins, specifically Al, Ca, Cu, Fe, Mg, Se, Zn, and vitamins B1, B12, E, and C, occur in both AD and mercury toxicity. Aluminum has been found to increase mercury's toxicity. Similar biochemical changes in AD occur with mercury, including changes in levels of homocysteine, arachidonic acid, DHEA sulfate, and glutathione.

Increased platelet activation, poor odor identification, hypertension, depression, and increased incidences of herpes virus infections occur in both mercury exposure and AD. Patients with AD exhibit higher levels of brain, blood, and tissue mercury.

This review of the literature concluded mercury may be an etiological factor in AD. Key Words: Alzheimer's disease; mercury toxicity; etiology

Introduction

The cause of Alzheimer's disease [AD] has been one of the great medical mysteries since Alois Alzheimer first identified this disease in 1906. It is the fourth leading cause of death in the United States affecting 4.2 million citizens. Although recent studies have shown a decline in incidence, it is still one of the most unmet challenges of our age [1]. The worldwide dementia prevalence is over 44 million people with an annual cost of $604 billion. It is estimated that the worldwide prevalence will triple to 135.5 million by 2050. At least 30 to 50 percent of all individuals above the age of 85 are affected by AD in the industrialized countries [2]. Approximately three to five percent of all cases have a genetic factor [3]. Many researchers believe that some environmental factor is responsible for the etiology of Alzheimer's disease.

The amount of neurofibrillary tangles found in affected Alzheimer brain regions correlates with the severity of the disease [3]. Some studies suggest that fibrillary nerve cell changes may begin to occur as early as 50 years before the onset of clinical symptoms. This would rule out old age as the cause. Neurofibrillary tangles in low amounts are found in about 20 percent of the population aged 20 to 30 years without clinical symptoms. By the age 70 to 80 years, 90 percent of individuals display neurofibrillary tangles in the brain [3]. Of this age group, 35 percent who have the highest number of histological detectable neurofibrillary tangles suffer clinically from symptoms of Alzheimer's disease.

Over the past three decades a number of studies have suggested a pathological role of inorganic mercury as being one of the causes of Alzheimer's disease [4] [5] [6]. This is underlined by the fact that mercury is ten times more toxic on neurons than lead, another neurotoxin [7]. Only mercury in low levels was able to induce the hallmark changes seen in AD compared to cadmium, manganese, aluminum, iron, and lead [6]. A substantial number of neurons, interneurons, and CNS-endothelial cells derived from people who die from a neuro-degenerative disease are full of intracellular mercury deposits, demonstrating that mercury easily penetrates the blood brain barrier, enters neurons, resulting in neuronal loss [8].

Mercury exposure has risen worldwide during the past decades from industrial activities, gold mining, medicinal usages, and burning of fossil fuels. Mercury released in the biosphere accumulates and is additive over thousands of years, and the content in the biosphere is today two to five fold over the pre-industrial era [9]. Mercury levels accumulate in organisms at the end of the food chain such as predatory fish and humans. Mercury content in tuna is rising 4 percent per year [10] and more than 100 to 1,000 fold over the past 1,000 years. Besides fish consumption, an important anthropogenic source of mercury originates from dental amalgams, which are comprised of about 50 percent mercury. The World Health Organization (WHO) stated in 1991 that the largest average daily intake of mercury in the general population originates from the dental amalgam [11]. Other medicinal sources are mercury based preservatives used in medications such as vaccines [12]. The World Health Organization has stated "there is no safe level of mercury." The purpose of this paper is not to identify the sources of mercury but to examine all known physiological changes that occur in Alzheimer's disease and determine if mercury could be causing these changes.

Methodology

To identify what physiological and pathological changes occur in Alzheimer's disease, a comprehensive AD book entitled *Encyclopedia of Alzheimer's Disease* [13], researched and written by Elaine Moore, was used as the primary reference for a summary of pathological and physiological changes that occur in Alzheimer's disease. Each of these AD changes was identified and then cross-referenced with mercury databases and PubMed to identify scientific articles that would explain whether, and if so how, mercury could cause these changes. Positive and negative reports were considered equally, and all data presented here helped explore the hypothesis. To our surprise, all 70 factors identified to date as occurring in Alzheimer's disease can be explained by mercury toxicity. The following is a brief summary of the physiological and pathological changes occurring in AD, followed by evidence describing how mercury could cause them.

Results

A. *HALLMARK BRAIN CHANGES IN ALZHEIMER'S DISEASE (AD)*

Three of the major diagnostic markers of AD, including neurofibrillary tangles, secretion of beta amyloid protein, and hyperphosphorylation of tau protein, can all be explained by mercury toxicity.

1. **Senile Plaques:** Neuritic plaques are composed of neuritic protein deposits found in excess amounts in AD brains. They are composed of beta amyloid protein [13].
   Mercury: Mercury stimulates the formation of beta-amyloid protein, which plays a role in the pathogenesis of AD by causing oxidative stress and neurotoxicity [14] [15] [16].

2. **Amyloid Precursor Protein (APP):** APP is the parent protein from which beta amyloid is derived. APP is broken down into fragments. A defect in APP may cause AD [13].
   Mercury: Inorganic mercury reduces the level of APP [17], and it strongly inhibits gamma secretase processing of APP [18].

3. **Tau Protein:** Accumulated phosphorylated tau protein is responsible for neurodegeneration in AD (13).
   Mercury: Mercury can significantly increase phosphorylation of tau protein [19] [20].

4. **Neurofibrillary Tangles (NFT):** NFTs are a characteristic brain finding in AD consisting of hyperphosphorylated tau protein [13].
   Mercury: Mercury increases the phosphorylation of tau protein resulting in NFTs [19].

5. **Memory Loss:** Loss of memory is a hallmark symptom of AD [13].

Mercury: Memory loss is the hallmark symptom of mercury toxicity [21].

B. NEUROTRANSMITTERS

Most major neurotransmitter functions are disturbed in AD. Likewise in mercury toxicity, the same neurotransmitters are affected.

1. **Acetylcholine:** Levels of acetylcholine are reduced in AD [13].
   Mercury: Acetylcholine is reduced in mercury toxicity [22].

2. **Serotonin:** AD brains have a deficit in serotonin [13].
   Mercury: Mercury inhibits the binding of serotonin to brain receptors [23].

3. **Dopamine:** Dopamine levels in the AD brain are low due to a deficiency of dopamine D2 receptors [13].
   Mercury: A negative correlation was found between total mercury and dopamine D2 receptors in the brain of wild mink [24].

4. **Glutamate:** Excess glutamate leads to neurodegeneration in AD [13].
   Mercury: Mercury inhibits glutamate uptake and stimulates release of glutamate [25].

5. **Nitric Oxide:** Excess nitric oxide can contribute to AD [13].
   Mercury: Mercury chloride induces nitric oxide synthetase [26].

6. **S-Adenosylmethione (SAMe):** AD patients with depression have decreased levels of SAMe [13].
   Mercury: Mercury inhibits SAMe [27].

7. **Norepinephrine:** AD brains have decreased levels of norepinephrine [13].

Mercury: Mercury decreases norepinephrine in brain synapses [28].

C. ENZYMES

Mercury's ability to interact with enzymes by its affinity to attach to sulfhydryl groups helps explain its toxicity.

1. **BACE 1 (Beta Amyloid Cleaving Enzyme):** BACE 1 has been implicated as the enzyme responsible for plaque in AD brains [13].
    Mercury: Methyl mercury increases end products of BACE 1 [29].

2. **Gamma Secretase:** Early onset AD patients have mutations to genes that produce gamma secretase which increases beta amyloid protein [13].
    Mercury: Mercury inhibits gamma secretase, suggesting the inhibition contributes to mercury-induced neuron toxicity [30].

3. **Kinases:** Kinases are protein enzymes responsible for phosphorylation of hydroxyl proteins. The enzyme is in high concentrations in AD [13].
    Mercury: Mercury induces hyperphosphorylation by activating kinase pathways [20].

4. **Cyclooxygenase-2 (Cox-2):** Cox-2 is released during inflammation and is over expressed in AD [13].
    Mercury: Mercury induces expression of Cox-2 [31].

5. **Cytochrome-c-oxidase:** Deficiencies of cytochrome-c-oxidase have been reported in AD patients [13].
    Mercury: Mercuric chloride inhibits cytochrome-c-oxidase [32].

6. **Monaamine Oxidase (MAO):** MAO forms hydrogen peroxide which is higher in AD [13].
   Mercury: Mercury increases MAO activity [33].

7. **Nitric Oxide Synthetase:** Elevated levels of nitric oxide synthetase contribute to AD [13].
   Mercury: Mercury induces nitric oxide synthetase [34].

8. **Acetyl Choline Transferase (CAT):** CAT is reduced up to 90 percent in AD [13].
   Mercury: Mercury inhibits CAT [35].

9. **Caspases:** In AD, there is activation of apoptosis caspases that underlie the pathology of AD [13].
   Mercury: Mercury increases activated caspases [36].

D. IMMUNITY

Alzheimer's disease is also known to be caused or aggravated by over activation of the immune system and inflammation.

1. **Complement:** AD is triggered by activation of complement 1 [13].
   Mercury: Studies show a correlation between mercury vapor and complement [37].

2. **Cytokines:** Cytokines contribute to inflammation. The beta amyloid in AD inflames the surrounding microglial cells [13].
   Mercury: Mercury significantly induces cytokine expression [38].

3. **Glial Fibrillary Acid Protein Antibodies (GFAP):** GFAP are autoantibodies associated with AD [13].
   Mercury: Methyl mercury induces GFAP [39].

4. **Interleukin 1 (IL-1):** Over expression of IL-1 in AD sets in motion the cytokine cycle [13].

Mercury: Mercuric chloride increases the release of IL-1 [40].

5. **Transforming Growth Factor Beta 1 (TGF B-1):** TGF B-1 regulates beta amyloid precursor protein synthesis. It is a cytokine that plays a central role in AD [13].
   Mercury: Thimerosal contains ethyl mercury which enhances the expression of TGF B-1 [41].

6. **TNF is Tumor Necrosis Factor (TNF).** Is a cytokine that contributes to AD. Excess levels of TNF are found in the cerebral fluid of AD [13].
   Mercury: Low levels of mercuric chloride increase the release of TNF [42].

7. **Beta-2 Microglobulin (B-2 M):** B-2 Microglobulin is a peptide that is increased in the amyloid disorder of AD [13].
   Mercury: B-2 Microglobulins are used as an early indicator of mercury toxicity in the kidney [43].

8. **Inflammation:** AD brains show evidence of inflammation adjacent to plaques [13].
   Mercury: Mercury exposure increases pro-inflammatory cytokines [42].

9. **Phosphodiesterase 4 (PDE 4):** PDE 4 is an enzyme that degrades cyclic amp (CAMP). CAMP underlies memory formation [13].
   Mercury: Mercury has been found to stimulate PDE 4 and reduce CAMP [44].

E. GENETIC

Studies suggest that AD has both a genetic component and environmental etiology.

1. **Apolipoprotein E4 (APOE4):** APOE4 may account for 60 percent of all AD cases. APOE 4 appears to promote the

binding of amyloid beta protein facilitating the formation of plaque [13].

Mercury: APOE4 has a reduced ability to bind mercury and thus potentiates mercury damage [3].

2. **Genetic Mutations:** Nearly 30 percent of early onset AD is linked to the presenilin 1 gene [13].
   Mercury: Mercury is known to cause gene mutations [45] [36].

3. **Presenilin Gene Mutation:** In the Antioguia region of Columbia, families with the Paisa presenilin 1 gene mutation develop AD nearly ten years earlier than individuals with the exact same mutation in Japan [46].
   Mercury: In the Antioguia region, individuals are exposed to the highest levels of mercury in the world from gold mines [47].

4. **Alpha-2 Macroglobulins (A2M):** A2M is a gene suspected of controlling the rate of beta amyloid protein production and is a possible susceptible gene for AD [13].
   Mercury: A2M levels were found to be significantly higher in mercury exposed workers [48].

5. **Hippocampus DNA Synthesis:** AD is characterized by death of cells in the hippocampus [13].
   Mercury: Mercury inhibits DNA synthesis by 44 percent in the hippocampus of rats [49].

F. MINERALS

1. **Aluminum:** Studies suggest an association between AD and aluminum [13].
   Mercury: Aluminum dramatically enhances mercury toxicity [50].

2. **Calcium:** Amyloid beta protein is increased in AD, which disturbs calcium concentrations and calcium metabolism [13].
   Mercury: Mercury increases intracellular calcium concentrations [51].

3. **Copper:** The Nun Study on AD found an association between high serum copper and AD [13].
   Mercury: Mercury can tie up metallothionein so the body cannot clear out toxic metals such as copper [52].

4. **Iron:** Increased levels of iron-binding protein melanotransferrin is seen in AD. Iron deposits are found in senile plaques. Beta amyloid production is increased in the presence of iron [13].
   Mercury: Iron levels in the blood have been correlated with mercury and memory [53].

5. **Magnesium:** Magnesium levels are depleted within the hippocampus of AD [13].
   Mercury: Mercury competes with magnesium and interferes with magnesium dependent pathways [53].

6. **Zinc:** Zinc tends to accumulate in areas of the brain most prone to AD damage. Zinc is suspected of aggregating beta amyloid deposits and pulls copper into the deposits. Zinc levels in the hippocampus are decreased in AD [13].
   Mercury: Mercury tends to displace zinc in energy producing pathways for neurotransmitters because they have similar chemical structure [54].

7. **Selenium:** Studies have shown that AD patients have lower levels of selenium in their plasma [55].
   Mercury: Selenium binds to mercury and neutralizes toxicity. Workers in mercury mines have lower levels of selenium in blood [56].

H. VITAMINS

1. **Folic Acid/Folate:** Low levels of folic acid increase the risk for AD [13].
   Mercury: Studies have found a negative correlation between serum folate and blood mercury [57].

2. **Thiamine (Vitamin B1):** Low levels of vitamin B1 increase the risk of developing AD [13].
   Mercury: Mercury reduces brain levels of vitamin B1 [58].

3. **Vitamin B12:** AD is characterized by vitamin B12 deficiencies [13].
   Mercury: Mercury can reduce the uptake of vitamin B12 [59].

4. **Vitamin C:** Evidence suggests that vitamin C (ascorbic acid) may prevent the onset of AD [13].
   Mercury: Workers exposed to mercury had lower levels of plasma ascorbic acid [60].

5. **Vitamin E:** Vitamin E deficiencies have been associated with AD [13].
   Mercury: Vitamin E provides complete protection from mercury toxicity in poisoned rats [61].

6. **Vitamin D:** Vitamin D may play a role in the pathophysiology of cognition decline in AD [13].
   Mercury: Little research has been done with mercury and vitamin D. Vitamin D deficiencies in AD may be explained by lack of sun exposure. Mercury causes inflammation and vitamin D is known to be anti inflammatory [62].

I. AMINO ACIDS, ANTIOXIDANTS, AND BIOCHEMICALS

1. **Homocysteine:** Homocysteine is elevated in AD and is toxic to the brain. It is associated with brain shrinkage [13].
    Mercury: Mercury toxicity overwhelms SAMe resulting in elevated homocysteine [63].

2. **Arachidonic Acid:** Arachidonic acid induces the polymerization of tau and induces apoptosis in neurons [13].
    Mercury: Methyl mercury induces the production of arachidonic acid of cerebellar granule cells [64].

3. **DHEA Sulfate:** The ratio of DHEA sulfate to cortisol is significantly lower in AD [13].
    Mercury: Mercury contributes to lower levels of DHEA [65].

4. **Glutathione:** Glutathione is essential for removing free radicals from the brain [13].
    Mercury: Rat studies have shown that mercury inhibits glutathione. In mercury toxicity, levels of glutathione decline and brain damage occurs [66].

5. **Antioxidants:** Free radicals are responsible for oxidative cell death associated with amyloid beta protein in AD [13].
    Mercury: Mercury produces free oxygen radicals that cause nerve damage [66].

6. **Hydrogen Peroxide:** Beta amyloid protein in AD increases the production of hydrogen peroxide [13].
    Mercury: The uptake of mercury vapor into cells is increased by hydrogen peroxide [67].

7. **Lipid Peroxidation:** The accumulation of lipid peroxidation products has been found in AD brains [13].
    Mercury: Mercury promotes lipid peroxidation [68].

8. **Glycosaminoglycans (GAGs):** GAGs are found in AD brain plaques. Sulfated GAGs have an affinity for beta amyloid [13]. *Mercury:* Rat studies have shown mercuric chloride increases serum GAGs [69].

9. **Brain Derived Neurotrophic Factor (BDNF):** BDNF is a cytokine that influences nerve growth in brains [13]. *Mercury:* Rat studies have shown methyl mercury-induced cell death occurs by way of receptors of BDNF [70].

10. **Melatonin:** Melatonin levels, which regulate sleep and wakefulness, are low in AD [13]. *Mercury:* Melatonin protects against mercury toxicity [71].

11. **High Density Lipoprotein (HDL):** HDL, the good cholesterol, reduces the risk of developing AD [13]. *Mercury:* Low HDL cholesterol is associated with high blood mercury [72].

J. OTHER FACTORS

1. **Platelets:** In AD, platelet activation rates are 30 to 50 percent higher compared to non-AD subjects [13]. *Mercury:* Normal platelet function is affected by high and low concentrations of mercury. Low mercury levels encourage clotting of red blood cells [73].

2. **Odor Identification:** Subjects who have mild cognition impairment and have poor odor identification scores are more likely to develop AD [13]. *Mercury:* Loss of smell is a symptom of mercury toxicity [74].

3. **Smoking:** Smoking increases the risk of AD [13]. *Mercury:* Dopamine levels are reduced in AD. Smoking increases neurotransmitters such as dopamine that can be seen as self medicating [75]. Animal studies have shown a negative correlation between total mercury and dopamine D2 receptors (24).

4. **Herpes Virus:** A high percentage of AD brains contain latent herpes simplex virus [13].
   Mercury: Mercury exposure increases herpes virus replication [76].

5. **Chlamydia Pneumonia:** Chlamydia bacterium has been found in parts of the brain of late onset AD [13].
   Mercury: Mercury has been found to increase chlamydia infections in animals [77].

6. **African Americans:** AD is very rare in west Africa. In the U.S., African Americans have an AD rate two times higher than whites [13].
   Mercury: In lakes across tropical Africa, fish have low levels of mercury [78].

7. **Depression:** Depression is more likely to occur in mild to moderate AD [13].
   Mercury: A major symptom of mercury toxicity is depression [79].

8. **Hypertension:** Chronic hypertension increases the risk for AD [13].
   Mercury: Serum mercury concentrations are associated with hypertension [80].

9. **Parkinson's Disease (PD):** Up to one third of PD patients develop AD [13].
   Mercury: Many epidemiological studies have shown an association between PD and exposure to mercury [81].

10. **Down Syndrome:** All patients with Down Syndrome will develop neuropathological hallmarks of AD [13].
    Mercury: An Egyptian study found elevated serum mercury levels in the Down Syndrome group compared to a control group, and they also had a significant increase in DNA damage [82].

Discussion

Mercury is a persistent bio-accumulative neurotoxic metal that can accumulate in the brain. Since industrialization, mercury in the air and water increased by 3 to 5 fold, and between 1977 and 2002 mercury increased in fish 4 to 5 fold [10]. Mercury in fish increases about 4 percent every year.

The half life of mercury in the brain is from several years to decades [83]. The World Health Organization rates mercury as one of the ten most dangerous chemicals to public health, and the U.S. Agency for Toxic Substances rates it in the top third. A study on autopsied brains by Weiner, et al, found that mercury concentrations in the occipital region of the brain increased with age [84]. Mercury exposure of humans is mainly derived from fish consumption, dental amalgams, and mercury based vaccines.

For decades there has been a controversy whether mercury exposure from these sources has clinical relevance. For fish consumption there are confusing observations regarding healthy outcomes because some kinds of fish are also big sources of selenium and omega 3 fatty acids which combat mercury toxicity. Additionally, the chemical form of mercury in fish already has reacted to molecules like cysteine or selenium. It may be much less toxic than the form which is used in toxicology studies or is produced in the gastrointestinal tract of humans by methylation of inorganic mercury from dental amalgam [85] [86]. Surprisingly, mercury vapor seems to be more toxic in rats than unreacted methyl mercury [87]. The question arises, which are the main sources of mercury in the human brain? Monitoring of blood, urine, or other biological specimens in living organisms has not correlated appropriately with brain mercury content.

Pendergrass exposed rats to mercury vapor and found average rat brain mercury concentrations increased significantly. The identical neural chemical lesions in the rat brain were similar or of greater magnitude as seen in AD brains. It was concluded that low levels of mercury vapor can inhibit polymerization of brain tubulin essential for formation of microtubules (88).

Blood mercury levels in patients (N=33) with AD were compared to a control group with major depression (N=45) and another control group (N=65) with non-psychiatric disorders. In early onset AD (N=13), blood levels were almost three fold higher than the controls. Blood mercury levels were more than two-fold higher in the whole AD group compared to the control group (89).

There appears to be a declining risk in Alzheimer's disease and dementia according to research by Langa (1). It was hypothesized this may be due to rising levels of education and successful treatment of cardiovascular risk factors. Since the 1980s, many dentists have curtailed or eliminated their use of mercury amalgams. The incidence of AD has declined in Sweden and Denmark, countries who have prohibited amalgams (1).

Evidence has been presented in this paper that explains how mercury can cause nearly all physiological and pathological changes that occur in AD. Seventy factors associated with AD were examined, and all can be explained by mercury toxicity.

Some of the studies referred to involve animals or were in vitro experiments, and one can't be sure if they are relevant to humans. However, the purpose of this paper was to show a relationship between mercury and pathogenic factors of AD, and this was the case.

Conclusion

The evidence suggests that mercury could be a factor in Alzheimer's disease. Mercury exposure has risen worldwide in the last decades. This is caused by the fact that mercury, which is released in the biosphere through industrial activities, medicinal usages, and burning of fossil fuels, is not able to degrade. Thus all mercury released into the biosphere will accumulate and become additive for thousands of years.

Typical brain damage in AD commences 20 to 50 years before symptoms manifest. Most of the U.S. population has been exposed to mercury from several sources beginning prenatal to postnatal. Further research is needed to explore the hypothesis that mercury may be an etiological factor in AD.

Table 1. Synopsis: Similar changes that occur in Alzheimer's Disease and Mercury Toxicity

A. Hallmark Changes

1. Senile Plaques - both increased;

2. Amyloid Precursor Protein - both decreased;

3. Tau Protein - both increased;

4. Neurofibrillary Tangles - both increased;

5. Memory - both decreased;

B. Neurotransmitters

1. Acetylcholine - both decreased;

2. Serotonin - both decreased;

3. Dopamine - both decreased;

4. Glutamate - both increased;

5. Nitric Oxide - both increased;

6. S Adenosylmethione (SAMe) - both decreased;

7. Norepinephrine - both decreased;

C. Enzymes

1. BACE 1 (Beta Amyloid Clearing Enzyme) - both increased;

2. Gamma Secretase - both decreased;

3. Kinases - both increased;

4. Cyclooxygenase-2 (Cox-2) - both decreased;

5. Cytochrome C Oxidase - both decreased;

6. Monaamine Oxidase (MAO) - both increased;

7. Nitric Oxide Synthetase (NOS) - both increased;

8. Acetyl Choline Transferase - both related;

9. Capases - both decreased:

D. Immunity

1. Complement - both increased;

2. Cytokines - both increased;

3. Glial Fibrillary Acid Protein Antibodies (GFAP) - both increased;

4. Interleukin 1 (IL-1) - both increased;

5. Transforming Growth Factor Beta (TGF B1) - both increased;

6. Tumor Necrosis Factor (TNF) - both increased;

7. Beta-2 Microglobulin - both increased;

8. Inflammation - both increased;

9. Phosphodiesterase 4 (PPF-4) - both decreased;

E. Genetic

1.   Apolipoprotien E4 - both related;

2.   Genetic Mutations - both related;

3.   Alpha 2 Macroglobulin (A2M) - both increased;

4.   Hippocampus DNA Synthesis - both related;

F. Minerals

1.   Aluminum - both related;

2.   Calcium - both increased;

3.   Copper - both related;

4.   Iron - both related;

5.   Magnesium - both decreased;

6.   Zinc- both related;

7.   Selenium - both decreased;

G. Vitamins

1.   Folic Acid - both decreased;

2.   Thiamine - both decreased;

3.   Vitamin B12 - both decreased;

4.   Vitamion C - both decreased;

5.   Vitamin E - both decreased;

6.   Vitamin D - both related;

H. Amino Acids, Antioxidants, and Biochemical Changes

1. Homocysteine - both increased;

2. Arachidonic Acid - both increased;

3. DHEA Sulfated- both decreased;

4. Glutathione - both decreased;

5. Antioxidants - both related;

6. Hydrogen Peroxide - both increased;

7. Lipid Peroxidation - both increased;

8. Glycoaminoglycans - both increased;

9. Brain Derived Neurotrophic Factor (BDNF) - both related;

10. Melatonin - both decreased;

11. High Density Lipoprotein - both decreased;

I. Other Factors

1. Platelets - both increased;

2. Odor Identification - both decreased;

3. Smoking - both increased;

4. Herpes Virus - both increased;

5. Chlamydia Pneumonia - both increased;

6. African Americans - both related;

7. Depression - both increased;

8. Hypertension - both increased;

9. Parkinson's Disease - both increased;

10. Downs Syndrome - both increased;

CHAPTER EIGHT

Two Pioneers

Providing Evidence That Mercury Causes Alzheimer's Disease

It has been difficult for scientists to secure funding for research regarding mercury toxicity and Alzheimer's disease (AD). Why this difficulty you might ask if there is so much evidence linking mercury and AD? Government funding from NIH in the United States to research amalgam mercury toxicity is nonexistent, as is funding in Europe. Scientific journals are reluctant to publish studies regarding mercury and health issues if dental amalgams are mentioned. For over 150 years people around the world have been treated with dental amalgams to fill cavities. The potential liability to the dental profession is enormous if it can be proven that dental amalgam is harmful, especially if it is causing AD. When I was doing research in the 1990s, it was thought the estimated cost to the dental profession would be one trillion dollars for the harm they have caused humanity. The liability probably has more than doubled since that time. My funding in the 1990s came from a private source. One might imagine the dental profession would do anything to keep the harmful effects of mercury under wraps. In Colorado, I found this to be true as did Dr. Hal Huggins, the dentist who first brought awareness to amalgam mercury poisoning. The dental profession, American Dental Association (ADA), dental schools, state governments, NIH, FDA the Alzheimer's Association, scientific journals, academia are all willing to let millions of people die from the harmful effects of dental amalgams in order to maintain economic status quo.

Fortunately, there have been pioneer scientists who have risked their careers to bring this knowledge to the forefront regarding amalgam mercury toxicity and Alzheimer's disease. Leading the way in the United States has been Boyd E. Hayley, Ph.D, the former department head of Chemistry at the University of Kentucky. In Europe the leading researcher is Joachim Mutter, M.D. who was on the faculty at the University of Frieburg Medical School in Germany.

Dr. Joachim Mutter

Through research and scientific articles, Dr. Mutter has done the most to raise consciousness this century in Europe about the relationship of mercury and AD. In 2010, Dr. Mutter, et al, had a scientific paper published in the Journal of Alzheimer's Disease entitled "Does Inorganic Mercury Play a Role in Alzheimer's Disease? A Systematic Review and an Integrated Molecular Mechanism," which provided evidence that mercury may be causing AD, which I will summarize in this chapter.

Dr. Mutter writes mercury (hydragyrium = Hg) is the most toxic non-radioactive element. About 80 percent of mercury vapor is absorbed when one breaths mercury vapor because of its uncharged monoatomic form. (Dental amalgam is constantly giving off mercury vapor where 80 percent is absorbed into the blood stream from the lungs.) It is highly diffusable, it is lipid soluble, and it easily crosses the blood brain barrier into lipid layers of cells and cell organelles such as mitochondria. It also penetrates the mucosa and connective tissue of the oral and nasal cavities and may be transported into nerve cells. It is then oxidized from the comparatively inactive Hg0 state to its ionic form of Hg++.

The mercuric ion reacts immediately with the intra cellular molecules such as enzymes, glutathione, tubulin, ion channels, and inhibits their activities thus interfering with normal cellular functions. Even very low mercury levels can decrease glutathione (GSH) levels and increase oxidative stress that leads to cytotoxicty. Its affinity for selenium disrupts the redox balance, especially in the brain.

Methyl mercury, when bound to a thiol group such as cysteine, can easily cross the blood brain barrier and is then transported into glial cells and neurons. Because of its charge, it can be trapped within cells and thus the brain. The brain is the major target organ for elemental gaseous mercury, Hg0. Once inside the brain, the half life for mercury is 22-27 years. Autopsies have shown mercury in the brain 17 years after exposure. (Remember, mercury vapor from amalgams is constantly being released.) In contrast, the half life of mercury in the body is around 30 to 60 days.

Humans take in about 2.4 ug of organic mercury per week if consuming one fish per week of which 2.3 ug is retained. The main source for intake of elemental mercury (Hg0) are dental amalgams. Fillings consist of 50 percent mercury which evaporates continuously in vapor form when in the mouth. Between 1.2 ug and 27.0 ug of mercury is taken up per day, and between 1.0 and 22 ug of mercury is retained. Variables that affect the amount of mercury vapor emitted from amalgams include the number, size, and age of the filling, as well as chewing habits and the influence of drinking hot liquids.

AD was first described in the scientific literature in 1907. Today between 15 and 50 percent of people 80 years and older suffer from AD. Currently about 24 million people worldwide suffer from dementia with the number projected to double every 20 years. By 2050, nearly 1 in 15 Americans are predicted to suffer from AD. In 1998, the chance of a 55 year old woman developing AD was 33 percent compared to 16 percent for men.

AD pathology results from a gradual build up of amyloid plaque resulting from amyloid-B being produced at a higher rate. These plaques induce inflammation and free oxygen radical production leading to neuroinflammation and neurodegeneration.

A second independent process involves hyper phosphorylation of tau protein, which leads to the breakdown of microtubules and the neurocytoskeleton. As a result, there is an accumulation of neurofibrillary tangles resulting in neuroinflammation and thus reinforcing the cycle.

In the brain, the degeneration process starts in the entorhinal cortex and the basal ganglia, especially in the nucleus basilis Meynert. Then it spreads to the hippocampus, and eventually it effects other parts of the cerebral cortex. Short term memory is the first to be noticeably affected.

Several genetic factors contribute to AD. Apolipoprotein E (APOE) genotype is a consistent factor. APOE is a transporter protein of cholesterol that may operate as a free-radical scavenger. The APOE 4 genotype confers up to a 15 fold risk increase compared to the APOE 3 genotype, while the APOE 2 genotype is protective.

All three APOE forms consist of 299 amino acids, and the only difference is that APOE 4 has an arginine in position 112 and 158, where APOE 2 has two cysteines, and APOE 3 has one arginine and one cysteine. Cysteine is a sulfhydryl which is capable of binding metals such as lead, copper, zinc, and mercury. It is believed that APOE 2 with its two cysteines is able to transport mercury out of the cells, and this may be why APOE 2 is protective against AD.

The APOE lipoprotein is taken up by the neurons via the APOE R2 receptor. Selenoprotein P (Sel P) which provides selenium for selenoprotein synthesis is also taken up by the APOE R2 receptor. The differential competition for uptake between APOE isoforms and Sel P might affect selenoprotein status and vulnerability to oxidative stress. Sel P is physically associated with AB plaque and neurofibrillary tangles in the AD brain, which suggests a role for impaired selenoprotein functioning in AD pathology.

Dr. Mutter did a scientific literature search relating AD, inorganic mercury, and neurotoxicity. Out of the 158 studies, 86 were included that met his standards for analysis. The meta analysis found:

1. A significant correlation between measure of cognitive functioning and mercury excretion for a mean excretion of 34 ug. There was a significant effect for differences in cognitive performance and measurements between mercury exposure and non-exposure for attention and mercury, that is, there was a dose-response relationship.

2. The mercury excretion in urine correlated with the measurement of cognitive functioning.

3. There was a relationship between strength of exposure (dentists vs personnel) markers of exposure and test results.

4. All the studies confirmed the toxic effects of mercury on neurons and neuronal tissue, which reproduces the pathological signs of AD.
   The dentists and their staff are exposed to mercury long term. All studies found a significant correlation between the level of mercury in blood, urine, nails, and hair.

5. Studies on health effects in people with amalgams have largely been negative because most people have had amalgams. The only study showing effects involved a young sample (22.44 years and 22.3 years respectively, the author's study) where the control group had never been exposed to amalgam mercury. (The amalgam group had significantly more health and mental health problems compared to the control group without amalgams in that study). All other studies investigated older people. People with no teeth left, usually older people, often did worse than those with teeth and amalgams. The people without teeth probably all had amalgams at one time.

Mutter then studied mercury exposure, accumulation, and excretion of mercury in AD patients. In a prospective cohort study, there was a negative correlation between mercury in nails and age or progression of dementia. More severely demented people do not excrete mercury as healthier people, probably due to the fact that their body is less able to excrete mercury.

A cross sectional study found significantly more mercury in plasma and non-significantly more in the cerebrospinal fluid of AD patients.

In a series of small studies, there was more mercury excreted in urine of AD patients than in age-matched controls and less mercury in blood of AD patients. The findings were not significant because of the small sample.

A retrospective cohort study found a probable exposure to mercury in 4.1 percent of 170 patients with AD and 2.4 percent likely exposure in controls. Another study found that the number of amalgams was not different in 66 AD patients compared to the control. Research has found AD patients had higher blood mercury levels that correlated with higher amyloid B levels in cerebrospinal fluid.

Four of nine autopsy studies document various changes in AD brains suggestive of mercury effects. One study found significant mercury in 81 brain samples of 14 AD patients compared to an age-matched control and more mercury in gray matter of AD brains compared with white matter. Mercury accumulates in the cerebellum, thalamus, putamen, and upper parietal and occipital lobe of AD brains.

A study of brain levels of mercury found significantly higher mercury levels in the amygdala, the nucleus basilis of Meynert, and non-significant levels in the hippocampus of 14 AD patients compared with an age matched control, while another study found significant higher mercury levels in microsomes of AD brains.

Higher brain mercury levels and lower mercury in nails were found in three AD patients compared to a control. There were four studies that found no significant difference or slightly and non-significant lower mercury levels in AD brains compared with a control.

The next phase in Dr. Mutter's research looked at animal studies and in vitro animal studies. Animal studies can easily be controlled for mercury and without worry if animals ever had amalgams. Five studies showed that rats exposed to mercury vapor had higher levels of brain mercury than the control. Two animal studies found that Parkinje cells of the cerebellum were specifically prone to accumulation of mercury following exposure.

In Vitro studies found:

1. Mercury interferes with polymerization of microtubules.

2. After mercury exposure, there was an increase of secretion of both I-40 and I-43 forms of amyloid B protein which promotes hyperphosphorylation of tau protein.

3. Mercury was found to change mitochondrial structures that induced a stress response in astrocytes, which interfered

with cell maturation such as DNA, glutathione levels, and microtubules.

4. Mercury was found to disturb the interaction between tubulin and GTP (quanosine 5' - triphosphate).

5. Amalgam mercury exposure is toxic for nerve cells in vitro.

6. Mercury interferes with membrane structure leading to axonal degeneration and neurofibrillary tangles (NFT).

Dr. Mutter writes that there are mixed results in studies involving AD and mercury. He emphasizes that any study that does not take into consideration if the person ever had amalgams is flawed. Remember the half life of mercury in the brain is about 22 years where it is entrapped. All these studies on humans did not take this into consideration and thus, the studies are flawed.

Animal studies are better for showing how mercury can be causing AD. In humans, there are conflicting results because most people have amalgams or had amalgams if they had dentures. Mutter writes, " there is no solid longitudinal evidence linking mercury toxicity with AD. At the moment, the evidence consists of pieces of the puzzle that are coherent and suggesting a causal relationship, but not absolutely compelling. The human studies also do not take into consideration whether the person has an APOE allele, which is important in removing mercury from the cell. Individual differences in detoxification capacity and genetic vulnerability make it difficult to run a perfectly controlled study". Dr. Hayley would say one also has to consider exposure to other metals that would exacerbate the harmful effects of mercury.

Mutter concludes that economic interests due to potential cost of litigation are driven for maintaining status quo. He emphasizes we need to become active to remove mercury from public and ecological circuits and replace it with less toxic alternatives. (The costs of dementia are overwhelming, let alone all the suffering from patients and their caregivers).

Is Dental Amalgam Safe for Humans?

This was the title of a scientific paper in 2011 written by Dr. Mutter in the *Journal of Occupational Medicine and Toxicology* in response to a claim made by The Scientific Committee on Emerging and Newly Identified Health Risks (SCNIHR) in a report to the European Union Commission. The report concluded that dental amalgam does not pose a health risk. Mutter found so many errors in the report that totally ignored scientific studies which indicated otherwise, and he was compelled to respond.

Mutter wrote that dental amalgam is the main source of tissue mercury. There is approximately a two- to five-fold increase in mercury levels in living individuals with dental amalgam as well a two- to twelve-fold increase in several body tissues observed in deceased individuals with dental amalgam. Animal studies confirm that dental amalgam leads to significant increased levels in the tissues. According to these studies, dental amalgam is responsible for at least 60-95 percent of mercury deposits in human tissue.

Studies have shown that mercury from dental amalgam is transformed into organic mercury by microorganisms in the gastrointestinal tract. A study by Leistervo, et al, found a 3 fold increase of methyl mercury level in saliva of individuals with amalgams compared to those without amalgams. Fish consumption was identical in both groups. The form of methyl mercury derived from dental amalgam may be up to 20 times more toxic than the form of methyl mercury found in fish.

Mercury vapor inhalation in doses that occur in humans with many amalgams led to pathological changes in the brains of animals after 14 days. One autopsy study found individuals with more than 12 amalgam fillings had more than 10 times higher mercury levels in several tissues, including the brain, compared to individuals with only 0-3 amalgam fillings. The average mercury level in the brain of European Union citizens with more than 12 amalgams was 300 ng Hg/g of brain tissue, which is well above mercury levels proven to be toxic in vitro on neurons (0.02-36 ng Hg/g).

In another autopsy study, individuals with more than 10 amalgams had 504 ng Hg/g in their kidney tissues compared to 54 ng Hg/g of those with 0-2 amalgams. The liver comparison was 83.3 to 17.68 ng Hg/g. Mercury levels in the thyroid and pituitary gland correlated with the number of amalgams.

About 20 percent of individuals in the age group of 20 years, 50 percent of individuals in the age group of 50 years, and 90 percent of people in the age group of 85 years living in Germany, showed pathological changes in their brains that are typical for both AD and mercury toxicity. About 80 to 90 percent of German people have dental amalgams over many decades. Between 30 to 50 percent of Germans above the age of 85 years have AD.

Maternal Amalgam

Maternal amalgam fillings lead to a significant increase in mercury levels of fetal and infant body tissue. Studies have found placental, fetal, and infant mercury correlates with the number of amalgam fillings of the mother, as does amniotic fluid and breast milk. Drasch, et al, found mercury levels of up to 20 ng Hg/g in German infant brains, mainly caused by dental amalgams of the mother. Mercury levels of 0.02 ng Hg/g can lead to degeneration of axons. Adams, et al, observed significant increases of mercury in baby teeth of infants with autism compared to a control. Epidemiological studies indicate that mercury exposure is responsible for autism or a deterioration of the disease. Exposure to maternal amalgams and thimerosal, the mercury preservative in vaccines, together with a genetic factor may trigger autism. In animal experiments, vaccination with thimerosal leads to symptoms similar to autism. Autistic children have decreased levels of the natural chelator glutathione.

One study found that autistic children who were exposed to mercury in their mothers's womb had up to 15 times lower mercury blood levels than normal. The lower the hair mercury levels in infants the highest was the severity of autism. Low levels of mercury in the blood and hair suggests the child has difficulty excreting mercury.

Mercury Toxicity

Mercury has been shown to be 10 times more toxic than lead in vitro studies. It is the most toxic non-radioactive element. Mercury vapor is one of the most toxic forms of mercury, with 80 percent of mercury being absorbed by the respiratory system. Mercury vapor from amalgams penetrates into tissue with great ease. Once inside the cell, it is oxidized to the Hg2+ form, which has a high affinity for thiol groups. Chelating agents like EDTA, which normally inhibits heavy metals, has no inhibitory effects on mercury toxicity.

Genotoxicant and Health Issues of Amalgam Mercury

Dental amalgam fillings have been found to cause DNA damage to human blood cells. Low levels of inorganic mercury leads to significant DNA damage in human tissue cells and lymphocytes. This effect can even trigger cancer. Significant elevated levels of mercury in hair have been observed in breast cancer patients. Deposition of mercury in body tissue is most likely bound to selenium, meaning selenium is no longer available for the body. Studies have also shown that dental amalgam can induce mercury resistant bacteria, which can lead to general antibiotic resistance in oral bacteria and other body sites.

Consistent low-dose mercury exposure has been considered a possible cause for certain auto-immune diseases such as multiple sclerosis, rheumatoid arthritis, and systemic erythomatosis. (All of these conditions have been shown to improve following amalgam removal.) This occurs with mercury exposure below mercury's safety limits.

Studies have suggested that mercury may cause hypertension and myocardial infarction. Significant mercury accumulation has been found in heart tissue. (A study by the author found subjects without amalgams had significantly lower blood pressure than subjects with amalgams.)

A number of studies have shown that mercury plays a pathogenic role in AD. Elemental mercury has induced Parkinson's disease (PD), and one study found significant elevated mercury in 13 of 14 patients with Parkinson's disease. Another study found a correlation between blood mercury and Parkinson's disease, while other research found significantly higher amalgam-mercury exposure in individuals with PD compared to a healthy control.

Adverse health affects are found in the dental profession. One study found 85 percent of dentists and dental technicians tested showed mercury related toxicity in both behavior and physiological parameters. Female dental assistants exposed to amalgams showed a higher rate of infertility. Women with amalgams have been found to have increase levels of mercury urine and a higher incidence of infertility.

A five-fold increase of mercury levels was found in the cerbrospinal fluid of multiple sclerosis patients. The prevalence of

MS has been shown to correlate with the prevalence of dental caries (cavities). A retrospective study on 20,000 military individuals revealed a significantly higher risk of MS for individuals with more dental amalgams.

Dr. Mutter was involved in research of an amytrophic lateral sclerosis patient (Lou Gehrig's disease). Normally ALS is a fatal disease. A Swedish woman with more than 34 amalgams, who suffered from ALS, recovered after amalgam removal. Experiments have shown mercury vapor to promote motor neuron diseases such as ALS. Mercury enhances glutamate toxicity in neurons, which is a factor in ALS. Case reports have shown a correlation between accidental mercury exposure and ALS.

Dr. Mutter writes about the importance of glutathione's relationship with mercury. Glutathione is the major natural chelator for heavy metals in the body due to sulfhydrl groups containing cysteines. Only when mercury is bound to glutathione or selenium is mercury capable of leaving the body by way of urine or bilary excretion. Impaired glutathione production causes higher retention of inorganic and organic mercury in the body. Mercury leads to lower glutathione by inhibiting glutamine synthetase.

The UNEP reports a 3 to 5 fold increase of mercury in the environment over the last 25 years. In the European Union, the usage of amalgam amounts to 720 tons yearly, with dentists being the second larger user of mercury in the EU. A Swedish study found there are 40 tons of mercury in dental amalgams of Swedish people and 2200 tons of mercury in dental amalgam in the EU. In the United States, there are 1000 tons in amalgams. Mutter emphasizes that if mercury is involved in the pathogenesis of AD, the disease may need up to 50 years to be diagnosed clinically.

This paper written by Dr. Mutter was in response to a report by the EU commission (the SCNIHR committee) that concluded dental amalgam does not pose a health risk. The SCNIHR report was written by 1 engineer, 4 dentists, 1 toxicologist, and 2 veterinarians. Mutter discuses the strategies of organized dentistry to influence science and politics over the decades. This strategy was also used by the tobacco industry.

Dr. Boyd E. Haley

One of the leading scientist in the United States regarding mercury toxicity of dental amalgam is Boyd Haley, Ph.D a professor emeritus from the University of Kentucky and former department head of chemistry. Since the 1980s, Dr. Hayley, has been doing research on dental amalgam toxicity. He believes that amalgam mercury is an etiological factor in AD. In 2007, Haley wrote a scientific paper entitled, "The relationship of the toxic effects of mercury to exacerbations of the medical condition classified as Alzheimer's disease." The following is a summary of the paper.

Hayley writes that AD is a disease of unknown etiology. It is not directly genetically inherited and that some external factor, such as a toxicant exposure or an infection, must be involved. In the United States, the rate of AD is very similar in rural and urban areas. The rate is similar in all of the states. If a toxicant is involved, it must be of a very personal nature, like something we eat or place in our bodies. The involvement of infection agents such a bacteria, virus, or yeasts, while possible, seems not to be a direct causal factor. The NIH has spent huge amounts of funds to find the cause of AD but have excluded mercury in their studies. To date, they have failed to identify any microbial factor. If an infective agent was involved, it would have been identified by now.

Focal infections caused by microbes are known to produce toxicants that inhibit enzymes. Toxins produced by these oral faculative anaerobics, can react to generate compounds like methylthio mercury that can be exceptionally toxic. The toxicant must be able to cause or exacerbate many of the biochemical abnormalities found in AD brains.

The first case of AD was identified in the literature in1907, roughly 60 to 70 years after the introduction of mercury amalgams. Therefore, AD is most likely a recent human disease caused by human-generated mercury toxins, which is an iatrogenic mercury- poisoning disease.

Mercury toxicity inhibits enzymes that are synergistically enhanced by exposure to other toxins such as lead or cadmium (smoking). Dr. Haley was the first to show that aluminum enhances the toxicity of mercury. Aluminum has long been associated with AD. It is known that the intake of milk also enhances the retention of mercury in the body. Current science does not know what the combined toxic effect of many toxicants is when they are present with mercury. Therefore one cannot identify a safe level of mercury. Thiol reactive toxicants such as mercury, lead, cadmium and certain organics can exacerbate mercury toxicity. However, mercury is the only toxicant that has been able to reproduce many of the biochemical abnormalities and hallmark pathology changes that occur in AD.

Chronic mercury exposure occurs in most humans as most of us have grams of mercury stored in our amalgams, which is continuously being released in mercury vapor form only inches from our brain. Therefore, it is reasonable to propose that exposure to mercury is one of the major toxic factors involved in early onset AD. Exposure to other toxicants can enhance the toxicity of mercury and hasten the onset of AD. Thus, the presence of synergistic toxicity factors and genetic susceptibility factors prevent any simple correlation of mercury exposure or tissue levels to the onset of induced illness.

Research Review

Dr. Haley's research in the 1980s found that brain nucleotide protein tubulin and creatine kinase (CK) showed a greatly diminished activity when exposed to mercury. Both tubulin and CK are proteins that bind nucleotides GTP (quanosine 5' - triphosphate) and ATDC (adenosine 5'- triphosphate). Haley also discovered that both tubulin and CK had diminished activity in AD brains compared to controls. Most AD brains had lost 80 percent of the viable tubulin. Many heavy metals interfere with tubulin viability but only Hg2+ mimicked the biochemical abnormalities found in tubulin of the AD brain. Low concentrations of Hg2+ can cause loss of tubulin. Rat studies found when they were exposed to mercury vapor, they lost between 41 and 74 percent of the GTP binding capability in their tubulin, demonstrating a similarity to abnormality in AD brains.

Studies on snails found mercury at low concentrations could strip axons of tubulin, leaving behind neural fibrils that represent the initial stage of neurofibrillary tangles, a hallmark dignostic marker of AD.

Creatine kinase (CK) has a very reactive cysteine in its ATP binding site and is over 95 percent inhibited in the AD brain. Tubulin and CK both have a very reactive sulfhydryl in or near their nucleotide binding site. By binding to this sulfhydryl group, mercury can inhibit the biological activity leading to the inhibition of interaction with GTP and ATP. Mercury has a very high affinity for sulfhydryl and is a potent inhibitor of biological activity of both proteases.

Excitotoxicity Hypothesis and Mercury

The "excitotoxic" amino acid hypothesis says that the excitotoxic amino amine acid glutamate builds up in brain tissue causing neuronal death and thus AD.

Studies have shown the activity of Hg2+ sensitive glutamine synthetase (GS) is elevated in the cerebrospinal fluid of AD and can be used for potential diagnosis of AD. Research has found a profound impairment of GS in AD brain tissue, and mercury is known to impair GS. Both the metabolism and transport of glutamate would be affected by exposure to mercury vapor. This would lead to the observed elevated glutamate, GS levels, and glutamate excitotoxicity.

The mercury inhibiting effect on tubulin, CK, and GS does not conclusively prove that mercury causes AD. It definitely proves that mercury exposure could at least exacerbate the clinical condition of AD.

Glutathione (GSH) is the major bio molecule involved in the natural removal of mercury from the body. As mercury slowly accumulates in the body, it weakens the body's natural defense against itself and all forms of heavy metal toxicity, thus increasing the overall oxidative stress. It is well documented that mercury increases oxidative stress.

The FDA refuses to evaluate the mercury vapor released by dental amalgam or to make the American Dental Association (ADA) and amalgam manufacturers to provide such information. There is enough mercury in amalgams to continue a low chronic level exposure for years. A single large amalgam filling contains 1 gram of mercury (1 million micrograms) and a loss of 10 mcg a day would be enough mercury for 100,000 days or 274 years of exposure. A small tenth of a gram of mercury amalgam would last 27 years. There is enough mercury in amalgams to provide toxic exposure for the life of most fillings. Dental amalgams are called 'silver fillings' and are approximately 50 percent mercury by weight. Haley's research found that the level of mercury released by 1 square cm of dental amalgam from a single spill filling (each filling was slightly less than 1 square cm) was between 4 to 22 micrograms per day.

Supporters of amalgam usage utilize estimates of mercury release based on urine and blood levels that are widely known to vary drastically with time and is unreliable. Blood, urine, and fecal mercury measurement misses all the mercury retained in the tissue. It has been found that people with low levels of blood and hair mercury are not excreting mercury but retaining it in their tissue. High hair mercury levels indicate the ability to excrete mercury and low mercury blood levels indicate an impaired ability to excrete it.

Amalgams and Body Burden of Mercury

A number of studies report increased tissue levels of mercury in subjects that correlated to the number of amalgams. The WHO scientific panel found ranges from 3 to 70 mcg/day with the bulk of the mercury originating from amalgam fillings. An NIH study was done on 1127 military personnel soldiers who averaged 20 amalgam surfaces with a range of 0 to 66 surfaces. Each 10 surfaces increased the urine mercury level by 1 microgram/liter. Each 10 amalgam surfaces increased the total exposure by at least 12 mcg. The NIH study found that individuals with an average number of amalgam fillings had about 4.5 times the urine mercury levels as controls without amalgams. The soldiers with over 49 surfaces averaged 8 times the urine mercury levels compared to the control without amalgams. The urine mercury levels dropped by a factor of 5 after amalgam removal. Mercury excreted in urine is not a measure of total exposure, because tissue absorbs much of the mercury taken into the body.

A number of studies have found subjects with AD have higher concentrations of mercury in the brain. A 1988 study by Thompson, et al, at the University of Kentucky reported an elevation of mercury in the AD brain with the nucleus basilis Meynert being the largest trace-element imbalance observed to date in AD brains. Not to be outdone, the University of Kentucky dental school did their own study on brain mercury. As reported in a dental trade journal of the American Dental Association, they concluded their results did not support the hypothesis that dental amalgam mercury contributes to brain mercury level or that it was a pathological factor in AD. They checked several regions of the brain but not the nucleus basilis Meynert. Dr. Hayley had a number of criticisms of the study and wrote that it lacked credibility because it was impossible to explain how amalgams can increase kidney, liver, blood, urine, and fecal mercury levels and not increase brain mercury levels.

A study by Hock, et al, found that in early onset AD the blood levels of mercury were almost three-fold higher than a control group, and the increase was unrelated to dental status suggesting an unidentified environmental source. The study did not take into account early amalgam exposure.

Studies supporting the hypothesis that only mercury is capable of causing the three widely accepted major pathological diagnostic hallmarks of AD, which are amyloid plaques, neurofibrillary tangles, and phosphorylated tau.

Decades of research by Dr. Hayley has led him to believe that mercury is a major contributor to AD.

CHAPTER NINE

It's All in Your Head

The Hal Huggins' Story

The Third Amalgam War originated from a dentist in Colorado Springs, CO named Hal Huggins, DDS, MS. The title of this chapter is the same name as a book written by Dr. Huggins describing his journey of discovering the harmful effects of amalgam mercury. He has done the most in the world to raise consciousness regarding this dreadful malady and has suffered the consequences from the dental establishment who are trying its best to preserve status quo. This chapter is based on his book.

Dr. Huggins journey began in 1973 at a dental conference where he was lecturing. He was introduced to the harmful effects of amalgam mercury by a dentist from Rio de Janeiro named Olympio Pinto. Dr. Pinto told Huggins that mercury escaped from the filling and caused health problems. Huggins refused to believe him and said, "Mercury does not come out of the filling." Dr. Pinto relayed evidence to the contrary. He told Huggins about his father, who also was a dentist and removed amalgams from a dying child with leukemia. Because the girl's gums were hurting, he removed the child's amalgams. Within a few days, the child recovered from leukemia. The child's physician could not believe it was from the amalgams. So Dr. Pinto Sr. replaced an amalgam, and the girl's leukemia recurred the next day. Of course he removed it again. Pinto told Huggins about his success with a number of ailments, including Hodgkin's disease and sickle-cell anemia, diseases that showed up after amalgams were introduced to Brazil. Still, Huggins refused to believe him and asked Pinto where he got this information.

Pinto said he was in the masters program at Georgetown University, and mercury toxicity was his thesis. He had compiled the largest bibliography on mercury toxicity that existed. The thesis was not published because the National Institute of Dental Research, under the National Institute of Health, found out about his research and forced the university to halt the research. Pinto was forced to stop and returned to Brazil but still had in possession his unpublished research, which he sent to Huggins. (The situation of Pinto reminded me of my situation at CSU, where I was forced out of the Ph.D. program when I tried to publish my research on amalgam mercury toxicity.) In 1976, Huggins published Dr. Pinto's material under the title *Mercury Poisoning in America.*

Three weeks after the Mexico City conference, Dr. Huggins was lecturing at a dental conference and mentioned Dr. Pinto's anecdote about the girl recovering from leukemia following amalgam removal and the anecdote about a Hodgkin's disease patient recovering following amalgam removal. A dentist in the group said he had also been diagnosed with Hodgkin's disease and had his amalgams removed twelve years ago. He also recovered from Hodgkin's disease. It was at that moment Huggins began the Third Amalgam War.

The War

Huggins had thought if mercury was really a health problem, the American Dental Association (ADA) would have discovered the harmful effects one hundred years ago. He also thought the Food and Drug Administration would not allow toxic mercury to be used. Congress would have certainly legislated sanctions against mercury in amalgams.

From 1973 through 1979, Huggins began to remove amalgams. He met with many failures, but had a success rate of 10 to 15 percent of people recovering from illness after amalgam removal. He felt discouraged at what he thought was a low percentage of patient recovery. If amalgam mercury was safe, why were

10 percent recovering he wondered? Huggins' early successes included chronic headaches, brain fog, and multiple sclerosis. Still Huggins was discouraged and felt both a financial drain and emotional drain. Many dental friendships evaporated as he was telling dentists about mercury toxicity. He actually had someone spit on him.

In 1979 things fell into place. He began to have a higher percentage of successes because he discovered the role of electrical current with amalgams. By removing amalgams with a negative current first, the white blood cell count increased and symptoms began to respond more frequently. Huggins also purchased a mercury vapor analyzer to measure the mercury vapor coming off the amalgam. He relayed his findings to the ADA, which was not impressed with his discovery.

Huggins began lecturing more and describing his findings to other dentists around the country. His office was getting 5,000 calls a month for information on finding a dentist who believed in amalgam mercury toxicity. Many new disorders were responding, such as depression, chest pains, and terminal diseases, the key being unexplained. Huggins' success went up to 50, to 60, and to 70 percent. Interest was expanding as Huggins lectured about his findings. He had cut back on his dental practice to lecture between 60 to 100 times a year. His practice was booked up 18 months in advance. Then came the supreme blow, as he calls it, when organized dentistry sent out an edict. If you have Dr. Huggins on your program, no continuing education credits will be given to the doctors in attendance for the entire program. Huggins was devastated and upset. He had been seeing terminal diseases go into spontaneous remission. The academic world was asking for controlled studies. Of course, it is unethical to remove amalgams and put them back in. Huggins vowed to let the world know about mercury's potential destruction of health.

Michael Ziff, D.D.S., a Florida dentist, attended one of Huggins' lectures. He made a statement saying, "Advocates of dental amalgam filling claim that the safety of those fillings has been scientifically proven and documented numerous times and that it has been continuously reaffirmed over the 150 years of its use." Ziff had studied the toxic effects of mercury and quoted *A Textbook of Oral Pathology* that stated, "A toxic reaction from absorption of mercury in dental amalgam has been reported on a number of occasions."

Dr. Ziff wrote to the American Dental Association (ADA) and requested the actual primary research establishing the safety of amalgam mercury in patients. The ADA sent one secondary research article and a form letter stating the ADA's position. Ziff then called the ADA specifically requesting the primary research establishing the safety of mercury amalgams. He was told that none were available, but was referred to the National Institute of Dental Research and was told there was plenty of research available investigating the potential risks to dentists from the use of mercury. There was no research available for patient risk, because K.O. Frykholm in 1957 'proved' that mercury did not present a risk to patient. Frykholm had used techniques and instruments now known to be inadequate. Ziff then researched the literature and documented over 600 references on mercury and dental amalgam and concluded that amalgam mercury constituted a potential health risk to patients.

According to Huggins, the ADA keeps on changing its story about how many people are sensitive to metals in their fillings. In 1984, the ADA stated in *Science Digest* that only five percent of the population was sensitive to metals in their fillings. That would mean 12 million people. The next month they revised that figure to one percent. By 1989 the figure dropped to less than a million, and in 1991 it was back up to three percent. There have been no studies to support these claims. Huggins has never seen a study that showed amalgams to be safe.

Amalgams And Disease

Dr. Huggins had seen so many diseases that responded to amalgam removal that he notified the medical establishment, hoping medicine would step in. He was told, "If dentistry has created a problem, it is up to dentistry to investigate and correct the problem." Huggins started recording the changes that amalgam mercury toxicity produced and categorized them into five categories.

I Neurological Disease (Both Motor and Sensory)

Some diseases and motor symptoms include tremors, seizures, multiple sclerosis, and amytrophic lateral sclerosis (ALS or Lou Gehrig's disease).

Sensory symptoms include symptoms of Alzheimer's disease, emotional disturbances, unexplained depression, anxiety, and unprovoked suicide thoughts.

II Immunological

The diseases that have responded to amalgam removal include systemic lupus erythematosis, scleroderma, and rheumatoid arthritis.

III Cardiovascular

Cardiovascular symptoms include unexplained heart pain, high and low blood pressure, tachycardia, and irregular heartbeat to name a few.

IV Collagen

The collagen category includes osteoarthritis.

IV Miscellaneous

This category contains many unexplained symptoms such as chronic fatigue, brain fog, digestive problems, and Chron's disease.

Multiple Sclerosis

Removing amalgams from MS patients was by far the biggest proportion of Huggins practice. In 1966, Dr. E. Baasch published in the *Archives of Neurology and Psychiatry* the epidemiological evidence that supported an environmental agent as being the cause of MS. He thought mercury from dental amalgams was the primary cause of MS. Baasch described the pathology and symptomology of MS that was consistent with mercury toxicity.

In 1983, Theodore Ingalls, MD reported in the *American Journal of Forensic Medicine and Pathology* that the seepage of mercury from root canals or amalgam fillings might lead to MS in middle age. He found a direct linear correlation between death rates from MS and number of decayed, missing, and filled teeth.

Ingalls himself had developed MS after his root-canal tooth was filled with a mercury filling at the root tip. Ingalls later died from MS.

The Multiple Sclerosis Society has been notified of the MS connection to amalgams but has been averse to the connection. Huggins said they have actively campaigned against investigating into the mercury connection. One of Huggins' patients was the former chairman of the Kansas MS Society who had MS. He did very well following amalgam removal to the point where he was playing three sets of tennis each day. The MS Society wanted him to resign from the society but he refused.

Alzheimer's Disease

Dr. Huggins states, "Clearly mercury removal affects these people (with AD), although not always within a few days. It will take weeks or months, although occasionally, I see significant changes within the two-week stay they have with me."

Huggins wrote about a minister's wife with AD who had not spoken a word for 14 years. Ten days after amalgam removal she mumbled something which shocked the minister who asked, "What did you say?" "Nothing," she shouted at him. The minister called Huggins several months afterwards and jokingly asked Huggins, "Can you put just one or two amalgams back in her? She hasn't stopped talking for three months. I can't get a word in edgewise."

Not all AD patients respond. Three months after removal it usually becomes obvious if an improvement is going to occur. Six months is better than three, nine months better than six. It continues to improve for a year. After two years following amalgam removal, an AD woman obtained a driver's license. At one time she was getting lost in her own kitchen. Another anecdote is about a dentist diagnosed with AD who couldn't find his office he had practiced in for forty years. He recovered after amalgam removal and went back to work. Of course he stopped using amalgams in his practice.

Huggins has found AD patients difficult to treat because each step is critical and the patient is not always cooperative. He emphasizes the safeguards such as blood tests, nutrition, supplements, sequential removal, rubber dam use on removal, compatible dental material, and avoidance of the 7-14-21 day immune cycle. It is important that the caregiver understand the importance of these safeguards. He knows of AD patients going to dentists that are not knowledgeable of proper removal, and the patient does not get better.

Amyotrophic Lateral Sclerosis (ALS)

ALS, also known as Lou Gehrig's disease, is similar to MS as it attacks the nervous system, but it is much more vicious. Often it attacks the voice before motor function of moving the hands, legs, and feet. Its duration from start to death is around three years, an exception was Stephen Hawking.

From 1973 until 1990, Huggins could not get a response in ALS patients with amalgam removal. In 1991, he discovered that when the periodontal ligament is left in the socket following amalgam removal, bone cells will naturally try to grow to connect with other bone cells. However, if the periodontal ligament is removed from the socket, the bone cells do not attempt to heal by growing to find other bone cells. The top of the socket seals over two to three millimeters with bone under it. The cavity remains lined with inflammation lymphocytes. Huggins discovered if this cavitation is opened and the periodontal ligament is removed, ALS patients respond. Their mobility improves, their voice improves, as does their attitude.

There have been at least five independent studies that have found chronic exposure to mercury as an etiologic factor of ALS. Autopsy studies on people who die from ALS have found the level of mercury in the brain correlates with the number of amalgams.

Dr. Joachim Mutter, M.D., a physician in Germany had a scientific paper published on how an ALS patient recovered following amalgam removal.

Parkinson's Disease (PD)

After discovering the cavitation connection to ALS, Huggins applied it to Parkinson's disease patients, who also responded beautifully as Huggins puts it. His mother died of Parkinson's disease so he was especially happy to help his Parkinson's patients. Dr. Mutter reports that some of his PD patient have recovered following mercury amalgam removal and detoxification.

Huggins the Scientist

Determined to find out why people were becoming ill from their amalgams, Dr. Huggins received a masters degree in immunology from the University of Colorado at Colorado Springs. Huggins developed a mercury immune reactive test to determine if people were sensitive to mercury. He then developed other immune reactive tests for other metals used by dentists. He tested over 3,500 patients for immune reactivity and found 90.2 percent were sensitive to mercury, 95.2 percent to copper, 94.04 percent to zinc, 66.86 percent to silver, and 62.15 percent to tin, all components of the dental amalgam. Often their sensitivity resulted in autoimmune reactions such as MS, ALS, and lupus. Other symptoms he commonly found were frequent urination, ringing in ears, metallic taste, suicidal thoughts, depression, numbness, memory problems, frequent leg cramps, facial twitches, joint pain, chest pain, tachycardia, and nervousness. Many patients had airborne allergies and food allergies, with many allergies correlating to the time when amalgams were placed in the mouth.

It was only when Huggins was able to measure electrical currents from the amalgam that his success ratio went up. He found the electrical current in the mouth was probably the greatest stressor but little was known about it. Huggins purchased an ammeter, a device for measuring an electrical current. An electrical path exists between the filling itself and the natural fluids in the tooth under the filling.

Huggins found there are two types of electrical activity on the surface of the filling. One is like a standard battery, with two different metals in an electrolyte that can conduct electricity and produce a current. This is called a bimetallic cell. The other type of electrical activity involves a cell that exists on a filling called a differential aeration cell. It refers to electrical activity that exists between saliva in different areas that contain different amounts of oxygen. An area of saliva closest to the filling can react with oxygen-rich saliva exposed to the teeth that produces an electrical current.

Huggins had two concerns about the current.

1.   It causes a chemical reaction that releases toxic mercury.

2.   The process that directs an electrical current into brain tissue, perhaps produces pathology in the brain.

With a dental amalgam, any of the constituent metals can be given off, but the largest amount is mercury in the form of vapor. Huggins measured mercury values ranging from 0 mcg/mm3 to more than double the OSHA maximum limit for exposure, which is 50 mcg/mm3. When examining the mercury content of a seven-year-old amalgam, Huggins found that it contained 36 percent mercury, not 50 percent, meaning 14 percent of it had escaped since it was put in. This suggests that mercury loss may be a function of the electrical current from amalgams.

Mercury is a highly reactive chemical that can form a carbon hydrogen compound called methyl mercury, which is especially toxic to the brain and nervous tissue. There is a bacteria in the mouth called streptococous mutans that also causes dental caries. This bacteria also methylates mercury.

In the beginning, Huggins was getting ten to fifteen percent success of patients responding to amalgam removal. He then discovered the role of electrical current. His success rate went up to 70 percent.

Dr. Huggins then discovered that blood profiles and symptoms improved when the negative current fillings were removed first. Electric current is a measurement of the speed of a chemical reaction. The higher the current, the more mercury is being released from the amalgam. The human saliva is a good electrolyte, and all the dissimilar metals in the oral cavity are galvanic cells that produce electricity. Corrosion of the amalgam is the result of galvanic actions.

Huggins found people with multiple sclerosis, epilepsy, and emotional disease had a lot of negative current fillings. They usually had six or more amalgams with negative currents, 15 percent had fewer than six amalgams with negative currents, and five percent had no negative current. If the negative fillings are removed first, the chance of patients improving are good. If all the fillings are removed from the quadrant with the highest negative charged fillings first, the patient will have a reasonable chance of getting better. After all the quadrants with negative fillings are removed, then the quadrants with the highest positive currents are next removed. This procedure is called sequential amalgam removal and has been an effective way of removing fillings.

When an amalgam has undergone heavy corrosion, there is a layer of corroded material on the outer surface. Often a probe can't get through to measure a current. About 80 percent of the highly corroded amalgams are negative.

Gold and amalgams should never be in the mouth at the same time, as the amount of electrical current increases tremendously, leading to severe problems and more mercury vapors are released. If one has gold and amalgams, all amalgams should be removed first and then a gold crowns should be placed if needed. One can have temporary plastic crowns placed until all the amalgams are removed. Gold crowns do have current but are seldom negative.

Dr. Huggins believed a patient's recovery is due to both the removal of electric current and removal of mercury.

The suicidal patient is extremely sensitive to changes in electrical current. Huggins recommends only half the amalgams in a quadrant be removed in the first appointment and that the patient be with someone for the next 24 hours. He recommends giving protamine zinc insulin (PZI) in the area of amalgam removal, which stimulates formation of new blood vessels.

Huggins discovered the 7-14-21 day immune cycle, which has proven to be a major factor in most all his cases. The placement of an amalgam becomes day one of a predictive cycle of events in which the immune system receives a challenge. After seven days, another immune challenge occurs, and again on day 21. This cycle has to do with white blood cells and their 21-day cycle. One should never do dental work on the same day of the following week.

Mercury Toxicity

Dr. Huggins identified a number of ways that amalgam mercury affects the body. With mercury's affinity for sulfhydryl groups, it is easy to understand how it can affect the body's biochemistry including enzymes and neurotransmitters.

I. Mercury kills cells by interfering with their ability to exchange oxygen, nutrients, and waste products through the cell membrane.

II. Mercury can destroy DNA, the genetic code, which prevents the body reproducing that cell again.

III. Mercury can cause an autoimmune reaction by embedding itself in a cell membrane that gives the appearance of 'nonself.' This triggers the immune system to destroy that specific cell. Huggins believes this is an etiologic factor in diabetes, multiple sclerosis, and lupus as examples.

IV. Mercury can interfere with nerve impulse transmission, causing organs to get wrong messages. This may be related to memory problems, numbness, and tingling.

V. Mercury can attach to hormones and deactivate them.

VI. Mercury can bind to hemoglobin and affect its ability to carry oxygen. This may cause fatique.

Early Warning Signs of Mercury Toxicity

Dr. Huggins emphasized that it is not the presence of mercury that determines the potential for damage, but a person's reaction to mercury. It takes very little mercury to cause a reaction if a person is sensitive to mercury. The blood level of mercury tends to be relatively constant and is related to the number of amalgams as does the brain content. Mercury attacks many systems of the body, and the diagnosis must be based on the number of changes.

The following is a list of chemical indicators of early warning signs of mercury toxicity. Diagnosis of mercury toxicity is a professional judgement.

1. White blood cell counts above 7,500 or below 4,500.

2. Hematocrit (per cent of packed red blood cells) above 50% or below 40%.

3. Lymphocyte count of above 2,800 or below 1,800.

4. A blood protein level above 7.5 mcg per 100 milliliters of serum.

5. A blood triglyceride level above 150 mg percent ml.

6. A blood urea nitrogen level above 18 or below 12 percent.

7. A hair nickle level above 1.5 parts per million (ppm).

8. A hair mercury level above 1.5 parts per million (ppm) or below 0.4 ppm.

9. A hair aluminum level above 15 ppm.

10. A hair manganese level below 0.3 ppm.

11. An immune reaction to aluminum, nickle, mercury, copper, or gold.

12. An oxyhemoglobin level below 55 percent saturation.

13. A carboxyhemoglobin level above 2.5 percent saturation.

14. T lymphocyte's (white blood cell) main function is to maintain immunoglobulin antibodies. When the cells are attacked, the count goes to below 2,000 per cubic millimeter (the normal range is 2,200 to 2,400).

15. A DNA analysis of lymphocytes, as mercury can cause DNA damage.

16. Presence of root-canal treated teeth.

17. Symptoms of low body temperature. Mercury-toxic patients frequently have a low body temperature because the thyroid

hormone is affected. Body temperature increases after amalgam removal. Chronic fatigue, brain fog, and short term memory loss also improve.

18. Presence of amalgam and gold.

19. More negative currents than plus currents.

Dr. Huggins emphasized that mercury toxicity cannot be treated successfully until all the amalgams are out of the mouth.

Recovery From Mercury Toxicity

The following is a summary outline recommended by Dr. Huggins to recover from amalgam mercury toxicity.

I. The Porphyrin Test

If one is uncertain that mercury is causing the problem, Huggins recommends having a urine porphyrin test. Mercury is the only heavy metal that causes blockage of the conversion of porphyrin into hemoglobin and ATP, which is the energy storage unit. The blockage causes the porphyrin to accumulate in the blood until it spills over into the urine. To find a dentist to perform the test, one can contact the Huggins Applied Healing office by calling (866) 948-4638, or info@drhuggins.com.

II. Electrical Current Reading

The next step is to have electrical readings on the amalgams. The higher negative reading causes more conversion of mercury into methyl mercury. People with leukemia and ALS respond best when fillings are removed from higher negative to lower plus, such as -12, -9, -2, +20, +10, +3. The dentist may be jumping from one quadrant to another when removing the amalgam. Neither of these diseases lead to a full recovery according to Dr.

Huggins. However, Dr. Joachim Mutter, a physician in Germany, had a patient with ALS who had their fillings removed and fully recovered after three years.

For most cases, Huggins recommends quadrant removal. Remove the quadrant with the highest negative reading first, then the next highest negative quadrant, etc.

III. Basic Testing

This involves a complete blood count, blood chemistry, hair analysis, and urine mercury testing. Huggins believes if one has no health problems and is doing it for prevention, these tests could be skipped.

IV. Dietary Supplements

Dietary supplementation should be based on the test results, at least one to two weeks before amalgam removal. Supplementation seems to condition the body to excrete mercury and help with tissue repair. Excretion of both methyl mercury and inorganic mercury tends to increase if supplementation commences at least several days before amalgam removal.

Huggins recommends:

1. Transmix, a multimineral form that appears to improve the efficiency of cell membranes.

2. Eater's Digest is a digestive enzyme.

3. Vitamin C should be given before and after amalgam removal but not in the morning of the appointment.

Transmix, Eater's Digest, and vitamin C are good for pretreatment and maintenance after amalgam removal.

V. Serum Compatibility

Metals are used in fillings, crowns, and implants. There are over 1,000 different metallic alloys to choose from and each company has its own formula. Huggins has tested over 750 materials and found no dental material that is completely safe for everyone. Even "safe plastics," as he has found over 50 percent of the population react to them. Huggins does serum compatibility testing on all his patients to determine acceptability before placing any dental material.

a. Gold

The higher the percentage of gold, the better. Metals have been added to gold in order to harden it. Huggins has found nine percent of 3,500 patients to have an immune reaction to gold.

b. Copper

Copper is often mixed with gold, and over 90 percent of the population reacts to it.

c. Amalgam

Huggins found that 90.2 percent of the population is sensitive to amalgam mercury. A high copper content in an amalgam will cause amalgams to emit mercury 50 times faster than a conventional amalgam,

d. Composite Fillings

Composite are made of ground glass powder mixed with a plastic binder. Huggins found composite materials are suitable for 50 percent of the population, and the other 50 percent have an immune reaction.

e. Nickel

Nickel is used in 90 percent of all crowns. It is known for its allergic properties, carcinogenicity, and toxicity. Its major use is as a base on to which porcelain is fixed. Most of the adverse reactions to nickel are similar to mercury toxicity, including neurological disturbances, emotional problems, and blood disorders, such as leukemia that are initiated by a nickel crown. Huggins now uses only gold partials.

Prior to amalgam removal, one should have a serum compatibility test to insure the replacement fillings will not cause more reactive damage than your amalgams. Tests are available through mercury-free dentists or call the Applied Healing Office to get further information (866)-948-4638 or go to info@drhuggins.com.

VI. Other Considerations

A. Don't have crowns permanently cemented for at least one month after amalgam removal, especially if symptoms are severe.

B. Do not use bases like Dycal.

C. Avoid other mercury exposures. Huggins had come across a book listing 4,000 commercial uses of mercury. Some of the most common exposures to mercury come through food, cosmetics, medication, household chemicals, gardening chemicals, and professions and industry. In his book, Huggins lists over a hundred common sources of mercury.

D. Avoid having all dental appointments on the same day of the week to avoid a 7-14-21 day immune cycle.

E. If suicidal thoughts are among the symptoms, be sure to use protamine zinc insulin in each quadrant of amalgams that is removed. Have someone be with the individual for the first 24 hours after treatment.

F. Huggins believes proper nutrition is the most important component in restoring health. Stay away from refined foods, sugar, and alcohol, and eat a lot of fruits and vegetables. He found that improper nutrition destroys more recoveries than anything else.

VIII. Choosing a Dentist

Be cautious in choosing a dentist for amalgam removal. Most dentists are not familiar with the proper procedure of amalgam removal. For mercury-free dentists, contact the International Academy of Oral Medicine and Toxicology to get a list of mercury-free dentists. Remember, in all states except California and Florida, dentists are threatened with loss of dental license if they mention to a patient that dental amalgam mercury might be a hazard.

Contact Information

1. For compatibility of dental materials contact Biocomp Laboratories at (800) 331-2303 or go to their website at www. Biocomp. Org.

2. For supplements, mercury detection units, books, and additional information, contact Huggins Applied Healing in Colorado Springs at (866) 948-4638 or go to their website at www.Hugginsappliedhealing.com or info@drhuggins.com.

3. For a list of mercury free-dentists contact the International Academy of Oral Medicine and Toxicology at www. IAOMT.Org.

Dr. Huggins passed away on November 29, 2014, at the age of 77 years. His obituary referred to him as the elder statesman of holistic dentistry. He had given 2,500 lectures in 47 states and 16 foreign countries regarding the harmful effects of dental amalgams. He founded TERF, the Toxic Element Research Foundation. Hopefully his pioneering work will not be forgotten. Thank you Dr. Huggins.

CHAPTER TEN

Beating Alzheimer's

How Tom Warren Did It

Tom Warren wrote a book entitled *Beating Alzheimer's*, describing his journey after being diagnosed with Alzheimer's disease (AD) and recovering from this dreadful diagnosis. After reading his book, I applied and received a research grant from the Wallace Research Foundation to study the relationship between dental amalgam mercury and AD. Our proposal was to remove amalgams from AD patients and do before and after testing. For the control group we planned to tell these AD patients that we were putting a sealant on their teeth to prevent mercury from escaping. Of course, there is no such sealant, but we needed a control.

We ran newspaper ads to recruit AD volunteers, but got minimal response. Thinking that Alzheimer's patients don't read newspapers and their adult children are too busy with their lives to get involved in such a project, we asked the Alzheimer's Association if they would help us recruit. The reply we received was the Alzheimer's Association did not believe mercury was causing AD, and therefore they would not cooperate with us. Wow! We were very surprised at their attitude and proceeded to use the funds to research amalgam mercury's relationship with mental health.

Tom Warren's Diagnosis

Tom Warren lived in western Washington, and on June 11, 1983, his family doctor sent him to a hospital in Olympia for a CAT (computer assisted tomography) scan. Tom, age 50, was told he had Alzheimer's disease and may have only seven years to live. He took the CAT scan results and test results to another doctor, who concurred that the scan revealed atrophy and without a doubt the diagnosis of Alzheimer's disease was accurate. In his book, Tom displayed photos of his CAT scan.

He was devastated with the AD diagnosis, and fear and sorrow overwhelmed him. Tom became so exhausted that sometimes he had to crawl up the stairs. His brain felt like it was burning, and he was in pain with severe headaches. His thinking evaporated as he describes it, and he had become a cantankerous grouchy person. He experienced a rapid diminishing capacity to remember names and couldn't remember his own telephone number. Conversations with people were difficult. Tom prayed to die after realizing what he was about to go through.

Tom then sought out a variety of healthcare specialists and began reading veraciously about AD. Within 30 seconds, he would forget what he had read, so he marked the book. He visited a number of alternative health doctors. He had heard about the Heidleberg Stomach Acid test and asked a physician to perform the test. The test came back saying his stomach was more neutral than water, which gave Tom hope, because it meant food was not being digested properly and perhaps his brain was starving to death. He had also heard about what Dr. Hal Huggins was doing with amalgam mercury and proceeded to have his amalgams removed. Nearly four years after his diagnosis, and years of agony, frustration, and depression, he had another CAT scan performed, which showed that the disease process had reversed itself. He had the results checked by several specialists who were shocked at his improvement. A neurologist refused to believe it, and several radiologist had confirmed there had been brain shrinkage, and the lesions were not present. Tom's AD had been reversed.

Tom said, "I am living proof that death is not the only alternative, and that his Alzheimer's disease was reversed." Now full of joy, he drove a car again, conversed with friends, and returned to work after an absence of many years.

The Amalgam Mercury Connection

How did Tom reverse his Alzheimer's disease? The missing piece to the puzzle was Dr. Huggins information regarding mercury poisoning from dental amalgams. Tom decided to have all his teeth removed by an oral surgeon, along with his amalgams. The oral surgeon also used Panolex dental x-ray equipment, which revealed six amalgam fragments, four within the gum tissue and two large pieces about one-half inch down in the jawbone. Fragments are pieces of amalgam that are not cleaned up during dental treatment and have worked their way into the oral cavity. One of Tom's fragments was no larger than the point of a pin. One reason some people don't recover after amalgam removal is that amalgam fragments remain in the oral cavity.

The new composite material that Tom tried before biocompatibility testing caused Tom to have a severe swelling reaction. He was then tested on a Voll Electroacupuncture machine that tested for biocompatibility of dental filling material. He believes that according to Dr. Huggins' recommendation, everyone should undergo serum biocompatibility before replacing their amalgams.

Two dentists Tom had initially visited refused to remove all his teeth, until he had accidently broken a very old bridge. He had a large silver-mercury amalgam beneath a gold crown. As soon as it was removed, Tom felt instantly relaxed within his brain directly above the bridge. The same sensation was repeated several weeks later on the opposite side of his brain when the largest amalgam fragment was surgically removed from the jaw bone.

Amalgam and copper traces were found under most of his gold crowns. Tom was allergic to the mercury that nearly killed him, and now it was just a minor nuisance. Tom had scabs and sores cased by the mercury and copper leaching from the amalgam. They quickly disappeared after amalgam removal.

Following amalgam removal, Tom's arthritis disappeared. He had painful hemorrhoids that had plagued him for 30 years, and they disappeared. His blood pressure dropped from 147/90 to 117/70. Tom's heart irregularities that he had since childhood vanished in an instant following removal of the amalgam fragments from his jaw bone.

Tom wrote in his book that if you have five minutes of lucid thought process, go for it, and have your fillings removed and follow his technique.

Tom told of other people's condition improving after using his approach, including Alzheimer's disease, schizophrenia, epilepsy, arthritis, and multiple sclerosis. He has received letters from Alzheimer's people who have improved following his technique. Tom tells of a schizophrenic patient who was unresponsive to every treatment for six years. Within six hours after amalgam removal, he experienced a big improvement. A friend of Tom's was in a state mental hospital for 35 years. A psychiatrist followed Tom's instructions, and Tom's friend recovered in less than three months.

Allergies

Tom believes that Alzheimer's disease is an allergic response to mercury. Research has found AD is an active chronic inflammation process. The brain swelling could be an allergic reaction to an antigen, perhaps mercury. Clinical ecologists have been successful in treating schizophrenia, which they call a cerebral allergy or brain fog. Tom writes that 60 percent of cerebral allergy patients recover if the condition is correctly diagnosed. When schizophrenia, Alzheimer's disease, and neurological problems have been approached as allergy related, tremendous progress has been made. Some physicians believe Alzheimer's disease is an autoimmune disease. Dr. Jim Hardy, DDS, would say it was the immune system attacking mercury in the brain.

Tom suggests there may be a defective gene that makes a person susceptible to cerebral allergies. Tom's mother was schizophrenic, his grandmother was senile, and his sister committed suicide. Tom had several mental breakdowns, and at one time was declared mentally ill. Tom also considered committing suicide.

In the beginning of Tom's AD treatment, Tom was unable to think clearly to cooperate in the healing process. His homeopathic physician helped Tom clear his thinking, restore his strength, and helped with his disposition. Tom had moved into a new apartment and experienced all kinds of symptoms, only to find out he was reacting to mercury in the paint, which is now outlawed.

Amalgam Removal

It took 11 years from the time Tom's thinking process started deteriorating until he discovered Dr. Huggins' protocol. He emphasized that one needs to read Dr. Huggins' book *It's All in Your Head* before you have your amalgams removed, and you need to go to a dentist that will follow the exact protocol. He emphasizes, you are responsible for your health.

After the amalgams are removed, Tom emphasizes that you will need to remove the mercury from the rest of your body. He suggests a combination of amino acids, vitamins, and minerals that can be purchased at the health-food store. Tom found it takes nine to ten months to get rid of the mercury. If the body does not respond, he recommends that the patient be rechecked for amalgam fragments. Also consider EDTA (ethyline diamine tetraacetic acid) chelation therapy, which will completely withdraw heavy metals.

Tom said the galvanic current reaction between his gold crowns and silver fillings was so strong that he could illuminate a flashlight. Of course he was kidding. Huggins believes that the brain is short-circuited from the oral galvanic current and may be responsible for 50 percent of neurological diseases in the patients he has seen. About 90 percent of epileptic patients that Huggins has treated improve following amalgam removal.

As mentioned earlier, many years before amalgam removal, Tom had planned to commit suicide. How many suicides originate from disorders caused by amalgam mercury? Tom found that exercise helped him immensely and considers that is what saved his life.

Tom paid a heavy price as a result of amalgam mercury toxicity. Fortunately he discovered the cause and has helped enlighten many people to the hazards of amalgam mercury.

CHAPTER ELEVEN

Mercury Free

Dr. James Hardy's Experience

Mercury is considered to be the most toxic element, much more toxic than lead or arsenic. Many health disorders have an unknown etiology that can easily be explained by mercury toxicity. Until the early 1980s, the American Dental Association steadfastly claimed that mercury did not escape from the amalgam, which contains 50 percent mercury. Today it is known that mercury vapor is continuously escaping from the dental amalgam, 24 hours a day, with 80 percent of the mercury vapor entering the body through the blood stream from the lungs. A Swedish mercury toxicity researcher Patrick Stortebecker, M.D., Ph.D. wrote in his book *Dental Caries as a Cause of Nervous Disorders* that "The fall of the Roman Empire was caused by the intoxication from lead used in the water pipes. Might similarly, the mercury from dental amalgam be the ruin of today's western civilization."

How can this be with all the brilliant scientists in the world? There are over 200 symptoms listed in the medical literature for mercury poisoning. Amalgams have 700,000 parts per million of mercury (ppm), yet the FDA takes tuna fish off the grocery shelves with only 1 ppm of mercury. Mercury is harmful to every system in the body. The brain and central nervous system are the major targets for mercury poisoning. Symptoms of mercury poisoning are almost always mis-diagnosed. Most people with amalgams have had them for most of their lives and have been receiving chronic mercury exposure from their dental amalgams, which is cumulative, so it builds up in the brain and other parts of the body. Dentist, Dr. James Hardy, writes in his book *Mercury Free*, "Mercury is the master element of disguises. It has no smell, no taste, and can enter the body as a vapor, liquid, or solid. Inside the body, it can change from one form to another, getting by the body's natural defense barrier." The sad thing is there is no useful test to determine early stages of mercury toxicity or tell what damage has been done. Sometimes it can take years or decades of low level mercury exposure before symptoms manifest, such as in Alzheimer's disease. Each person has a different tolerance to mercury. This chapter is based on Dr. Hardy's book

Mercury Free

The excellent book entitled *Mercury Free*, written by dentist Dr. James Hardy, appeals to the lay person concerned about the safety of amalgam mercury. Dr. Hardy sheds light on disorders that respond to amalgam removal. Hardy strongly disputes the claims by the American Dental Association (ADA) that mercury amalgams are safe. He did a Medline computer search on the word mercury. In all his research he found no positive news about mercury. He emphasizes that mercury poisoning is often misdiagnosed and is harmful to every system in the body.

Hardy quotes an article by Dr. Bernard Tillis published in the ADA, 1984 August newsletter that reports, "Cardiovascular disease, neuromuscular disturbances, gastric intestinal disorders, respiratory and dermatological complications occur more frequently among dentists than in the average individual." The dental profession has the highest suicide rate of any profession. Could all this be caused by exposure to mercury? The answer is probably yes.

Birth Defects

Dr. Hardy was taught in dental school not to place mercury amalgams in pregnant women because of possible toxic mercury effects on the fetus. There are many women who do not know if they are pregnant and have mercury fillings placed in their teeth each year. Mercury in all its forms can enter the fetus and induce numerous birth defects by altering chromosomes and changing the shape of the DNA helix. Birth defects caused by mercury include cleft palates, cerebral palsy, mental retardation, as witnessed in babies in the Minamato Bay mercury poisoning. Other disorders are convulsions, speech disorders, and small heads.

High Blood Pressure

Hardy writes about mercury vapor from amalgams increasing blood pressure and wonders if this is the cause of hypertension that develops later in adulthood. High blood pressure can increase the risk of heart attacks and strokes. In our studies, we found subjects with amalgams had significantly higher blood pressure than subjects without amalgams.

Diabetes

According to Dr. Hardy, mercury inactivates the very enzyme that allows insulin to do its job. He believes that after mercury exposure, the patient may be left with plenty of insulin in the blood stream, but it may be inactive. Dr. Hardy thinks long term exposure from amalgam mercury may be a cause of diabetes.

Insulin has three sulfur bridges in its structure, which attracts mercury that may disrupt the sulfur bridge and thereby turn active insulin into inactive insulin as seen in diabetes mellitus. Diabetes mellitus is characterized by high blood sugar levels resulting from either low insulin production or inactive insulin. Mercury may interfere with insulin by changing its shape.

Abortions

Long term mercury exposure in animal studies has shown diminished sexual activity, lower sperm count, and lower fertility. A 1980 study found women exposed to mercury vapor had an increased rate of pregnancy problems and labor complications compared to women who had not been exposed to mercury. A study in Denmark found an increased rate of spontaneous abortions in a group of dental assistants working with mercury. A Polish study also found dental workers had an increased rate of spontaneous abortions. In a 1991 rat study, fathers were exposed to metallic mercury, and the rate of spontaneous abortions in rats increased significantly.

Anorexia and Colitis

Mercury exposure can cause anorexia, colitis, abdominal cramps, diarrhea, nausea and vomiting. A 1976 report describes how a teenage girl within two weeks after mercury vapor exposure exhibited the symptoms of anorexia, and symptoms of abdominal cramps, diarrhea, and painful bleeding.

In 1992, the Center for Disease Control reported that most fillings are placed in children between the ages of 9 and 15. Hardy believes this may be linked to anorexia in teenagers.

Abdominal pain and colitis are both symptoms of mercury poisoning. Amalgams should not be used in patients with colitis.

Colitis is also known as irritable bowel syndrome (IBS,) which is inflammation of the intestines. IBS represents about 50 percent of all gastrointestinal complaints. No anatomical cause for IBS has been found. Mercury exposure can worsen IBS. Hardy believes that eating foods which heat the amalgam can increase mercury vapor being released.

In Dr. Hardy's practice, he has removed amalgams in patients with colitis. After removal, there is a temporary increase in mercury vapor and symptoms flare up. The symptoms are less in every subsequent visit, and in every case the patient has dramatically fewer colitis flare ups. He replaces the amalgams with composite resins. Most cases of IBS/colitis cases begin between the ages of 15 and 30. Crohn's disease is similar to IBS.

Red Blood Cells

Mercury can attach itself to sulfur groups on the surface of red blood cells and cause peroxidation by free radicals, which lowers the red blood cell's ability to carry oxygen and carbon dioxide. This disturbance in the red blood cell may result in fatigue, a common symptom of mercury exposure. Our studies found subjects with amalgams had significantly fewer red blood cells compared to subjects without amalgams.

T Cells

T cells are white blood cells that have been activated by the thymus gland. They handle autoimmune disorders, cancer, viruses, bacteria, yeast, fungus, and other infections. The ideal percentage of T cells found in the white blood cell population is between 70 and 80 percent. In our studies we found subjects without amalgams had a significantly higher total T cell count than those with amalgams. Dr. Hardy tells of a 35-year old patient with advanced MS who had nine amalgams removed. Before removal, the T cell percentage was 60 percent. After removal, the T cell count rose to 81 percent.

An article published in the *Journal of Prosthetic Dentistry* tells of a 21-year old patient in good health who had six amalgams removed. Before removal, the T cell percentage was 47%. After removal the T cells rose to 73%, a 55% increase. Four amalgams were put back into the patient and the T cells decreased to 55%. The four amalgams were again removed, and her T cells rose to 72%. In our studies, we found subjects without amalgams had significantly higher T lymphocytes than subjects with amalgams.

Immunoglobulin G (IgG)

Dr. Hardy writes about two 1990 studies that found IgG levels decreased significantly in workers exposed to mercury vapor compared to non-exposed workers, and there was a significant decrease in IgA levels as well. IgG is the major defense against toxins, viruses, and bacteria. It gives immunity to the unborn fetus. One of our studies found subjects with amalgams demonstrated lower levels of IgG than non-amalgam subjects (p=0.13), and MS subjects with amalgams had significantly lower IgG than MS subjects who had their amalgams removed (p=0.061).

Hardy believes mercury may be responsible for antibiotic resistant infections. An article in *Dentistry Today* (May, 1993) reported that mercury leaching from amalgam fillings appears to encounter growth of bacteria that are resistant to antibiotics as well as mercury.

Sick Days

A 1991 study followed 383 people for nearly three years and found a strong link between the presence of amalgams and extra work days missed. The study found a 30% drop in sick days two years after amalgam removal.

Rheumatoid Arthritis (RA)

RA is a chronic systemic inflammation disease of the joints, considered to be an autoimmune disease. The body suddenly recognizes the connective tissue as abnormal and attacks it, causing inflammation. Dr. Hardy believes the label of autoimmune is incorrect, because the joints and outer sheath of nerves are largely made up of connective tissue called collagen. Two collagen molecules are connected by sulfur atoms, which mercury has a great affinity for. When mercury attaches to the sulfur of collagen, it initiates inflammation against the foreign collagen with mercury atoms. This also happen with MS as the immune system attacks the nerve collagen making up the myelin sheath that insulates the nerve, resulting in nerve demyelation of MS patients.

Many of Dr. Hardy's patients find relief from arthritis when they become mercury free. The best successes Hardy found in his practice with amalgam removal is rheumatoid arthritis and headaches. Nearly all rheumatoid arthritis patients showed improvement, and most recover dramatically and completely.

Cancer

Dr. Hardy believes there is a connection between cancer and amalgam mercury. A 1987 report linked a mercury-based fungicide with leukemia in farmers as well as their cattle.

Dr. Huggins reported of several individuals who recovered from Hodgkin's disease following amalgam removal. Patrick Stortebecker, M.D., Ph.D., reports of malignant gliomas of the brain being linked to amalgam mercury. There exists a correlation between a incidence of malignant gliomas and a high incidence of dental caries. In Sweden, there is a low incidence of glioblastoma in people that are edentulous (without teeth). Studies have revealed a high incidence of malignant gliomas of the brain among dentists and dental personnel.

Amytrophic Lateral Sclerosis (ALS)

Hardy also believes in a connection between ALS and amalgam mercury. Mercury intoxication resembles ALS. All mercury toxicity symptoms subside three months after mercury exposure subsided. Huggins reports of his success in treating ALS subjects with amalgam removal. Dr. Joachim Mutter, M.D. had a scientific article published on the recovery of an ALS patient following amalgam removal.

Attention Deficit Disorders

Hardy suggests attention deficit disorders (ADD) may be associated with amalgam mercury, the most common filling used in children. Symptoms of ADD are neuroses, anxiety, irritability, shyness, forgetfulness, tremors, headaches, nightmares and fatigue, all symptoms of mercury poisoning. The mercury preservative thimerasol used in vaccines has also been linked to ADD.

Gum Disease

Gum disease is considered the most common disease in America, with 85% of adults over the age of 35 having periodontal disease. It is a silent disease caused by a build up of plaque between teeth resulting from bacteria that may include bleeding gums and bad breath. The flow of waste from the plaque is called gingivitis. The bacteria in the waste flows onto the gums and causes inflammation. Gum pockets develop, and gum disease often leads to tooth loss.

A 1990 study found that subjects who were hypersensitive to mercury as determined by a patch test, had stomatitis (inflammation of the mouth) where ever the amalgam touched the gums. The inflammation disappeared when the amalgams were removed. Dr. Hardy strongly suggests that amalgams not be used in patients with gum disease.

Kidney Disease

Mercury has an affinity for the kidney. Six sheep in a Canadian study received 12 amalgam fillings. Within two months, the sheep lost 80% of their kidney function. The control sheep that did not receive mercury amalgams had no decrease in kidney function. Kidneys have a high capacity to concentrate mercury. Symptoms may vary from increased protein in the urine to complete kidney failure.

Tinnitus

Dr. Hardy had a 35-year old patient with tinnitus (ringing in the ears). The patient had all his fillings placed between the age of 9 and 12, and by the age of 12 he suffered tinnitus in his ears 24 hours a day. When he had his last amalgam replaced with a composite, the ringing disappeared.

DR. HARDY'S PROTOCOL FOR MERCURY DETOXIFICATION

Dr. Hardy recommends working with a medical physician, naturopath, homeopath, chiropractor, or other health care provider who is knowledgeable and qualified in detoxifying the body of mercury.

He has a four week mercury detox protocol following amalgam removal.

Week One

1. Vitamin C time release, 1000 mg, 2X daily
2. Vitamin A, 5,000 IU, daily
3. Vitamin E, 200 IU, daily
4. Vitamin B, 15 to 20 mg of each vitamin B, daily
5. One a day multiple vitamin
6. Chelated Zinc, 15 to 30 mg, daily
7. Manganese, 2.5mg to 5mg, daily
8. Magnesium, 450 mg, daily
9. Chromium, 0.2mg, daily

Week Two

1. Decrease vitamin C to 1,500 mg, daily
2. Keep taking plural supplements

Week Three

1. Decrease vitamin C to 1,000 mg, daily
2. Decrease vitamin A to 2,500 IU, daily
3. Keep taking all other supplements

Week Four

1. Decrease vitamin C to 500mg, daily
2. Keep taking all other supplement

Dr. Hardy recommends stopping the detox program at the end of week four. He suggests to his patients to avoid sugar, carbonated beverages, caffeine, tobacco, excess dairy or meat products, and processed foods.

The diet should consist of complex carbohydrates, green salads, fresh vegetables, fresh fruit, and protein sources other than red meat. Do not use megadoses of vitamins before consulting with your physician.

CONCLUSION

The World Health Organization (WHO) has concluded there is no safe minimum dose of mercury. Symptoms are known to occur at least among some of the population at every level of exposure. The WHO states that dental amalgam mercury is the leading contributor of mercury exposure to the population. This was confirmed by mercury researchers Drs. Clarkson, Hursch, Nylander, and Friberg, who have concluded that dental amalgam was the largest source of inorganic mercury and mercury vapor in the general population. Mercury poisoning creates symptoms similar to lead poisoning that brought down the Roman Empire. Since the publicity about amalgam mercury toxicity began in the 1980s, the number of dental amalgams inserted into dental caries in the United States has decreased from almost 160 million in 1979 to about 100 million in 1990. Part of the reason is the increase use of composite resin fillings. Hopefully this trend will continue.

On February 18, 1994, the Swedish government banned mercury amalgams as a dental filling material. In 1994, Germany advised against using amalgams in all women of child bearing age. Austria, Denmark, and Finland planned to stop the use of mercury amalgam by the year 2000. The Alzheimer's disease rate has fallen in Sweden and Denmark. Perhaps this toxic pandemic will come to an end.

CHAPTER TWELVE

Preventing Alzheimer's Disease

Reducing the Risk

When I first started this book project regarding Alzheimer's diseases (AD), my primary goal was to gather the available information on how to prevent this terrible malady. It was during this process that I could prove that mercury was causing AD, and that the mercury primarily originated from silver dental fillings called amalgams. In the meantime, I had gathered numerous articles published in books, magazines, newspapers and the internet reporting about scientific studies that reduced the risk of developing AD. Most of this chapter is based on that literature. As you will discover, factors such as diet, exercise, mental health, and lifestyle all play a major role in preventing and delaying AD and dementia. Many of these factors can be related to mercury because of their antioxidant effects, their ability to rid the body of toxins, their anti- inflammatory properties, and their ability to neutralize the effects of toxins such as mercury. Science has shown that approximately 50 percent of AD can be prevented by following these preventive guidelines.

Dental Amalgams

Most people in the industrial world have silver dental fillings comprised of 50 percent mercury. What should you do if you have dental amalgams and no symptoms of dementia?

1. If you need a filling for a dental carie (cavity), tell your dentist you do not want a silver filling but a composite (white filling).
2. If you can afford it, have your silver amalgams removed. Go to a mercury-free dentist who has been trained in removing amalgams. There is a certain protocol in removing amalgams. Contact the International Academy of Oral Medicine and Toxicology at IAOMT.org for a list of mercury-free dentists in your area. They should know the proper protocol.
3. If you have gold in your mouth, it is wise to have your amalgams removed. Gold and silver amalgam set up an electrical current that releases mercury from the amalgam even faster.

The International Academy of Oral Medicine and Toxicology

In 1984, thirteen dentists founded the International Academy of Oral Medicine and Toxicology ((IAOMT) after attending a seminar on dental amalgam mercury toxicity. They were very concerned because of the lack of research in the scientific literature regarding amalgam mercury and it's safety. Within two decades, the IAOMT grew to 700 active members in North America with affiliated chapters in fourteen countries.

After years of research, the IAOMT has proven beyond a reasonable doubt that dental amalgam is a source of significant mercury exposure and is a hazard to health. It has taken the lead to educate dentists and allied professionals in the method of safely dealing with amalgam fillings and safely disposing of the waste.

The fundamental mission of IAOMT is to promote the health of the public at large. They have a list of mercury free dentists. Their website is IAOMT.org.

Diet

Diet is one of the most important factors in preventing the onset of dementia and AD. The diet that protects against heart disease also protects against Alzheimer's disease. It is a diet that contains many antioxidants that neutralize free radicals produced by toxins such as mercury.

Mind Diet

The Mind Diet is currently the most recommended diet to prevent AD. The diet was developed at Rush University Medical Center in Chicago and Harvard School of Public Health. Mind Diet is short for the Mediterranean Diet and Dash Diet (Intervention for Neurodegenerative Delay). Both the Mediterranean and Dash diets are healthy heart diets. One study found the Mind Diet reduced the risk of developing AD by 53 percent. The diet consists of ten optimal key food groups that include:

1. Green Leafy Vegetables: Six or more servings per week.

2. Other Vegetables: One or more servings per week.

3. Nuts: Five or more servings per week.

4. Berries: Two or more servings per week.

5. Beans: Three or more meals per week.

6. Whole Grains: Three or more servings per day.

7. Fish: One or more meals per week.

8. Poultry: Two or more meals per week.

9. Olive Oil: Primary oil for food preparation.

10. Wine: One glass per day.

Foods To Limit.

1. Red Meat: Less than four meals per week.

2. Butter and Stick Margarine.

3. Cheese: Less than one time per week.

4. Pastries/Sweets: Less than five servings per week.

5. Fried/Fast Foods: Less than one time per week.

The Mind Diet takes years off the brain. One study followed 960 older adults for 9 years. The mind seemed 7.5 years younger for those who followed the Mind Diet compared to those who did not.

Both the Mediterranean and Dash diets reduce the risk for heart disease, high blood pressure, and LDL cholesterol, the bad cholesterol.

Other Dietary Evidence

Foods that increase brain health include asparagus, which is full of vitamin A, split peas, legumes that improve mental processing, and walnuts that are a good source of alpha linolenic acid and omega 3 fatty acids necessary for good brain health.

A ten-year study of 950 older adults who ate daily just one or two servings of leafy green vegetables such as spinach and kale, had cognitive levels of people 11 years younger. Other studies have shown that people who consume more than four daily servings of vegetables have a 40 percent lower rate of cognitive decline compared to those who got less than one daily serving.

Olive oil is said to guarantee to boost memory power. A study of nearly 300 seniors found those who consumed at least five tablespoons of olive oil a day tested the best on memory and problem solving skills. Olive oil has vitamin E and other antioxidants.

Studies have shown that pomegranate juice reduces the level of beta amyloid protein that is found in the AD brain.

A study on adults aged 65 and older, who frequently ate at least five foods, such as vegetables, fruits, whole grains, legumes, olive oil, fish, and consumed moderate amounts of wine, and low amounts of dairy, meat, and poultry were found to have larger brain volumes than those who did not eat this way. The brain volume was comparable to about five fewer years of brain aging.

A 1983 study found subjects who ate meat, including poultry and fish, were nearly three times more likely to become demented compared to vegetarians. A 1990 study from Loma Linda University School of Medicine confirmed that vegetarians have less dementia than meat eaters. Other studies have shown that diets high in animal proteins have a higher correlation of AD prevalence, about twice the risk.

Vitamins

Many vitamins, such as vitamin E and vitamin C, are antioxidants that neutralize free radicals that can harm nerve tissue. Mercury produces free radicals that can harm brain tissue. One study conducted on 8,000 men and women who ate foods rich in vitamin E and C were less likely to develop AD six years later. A study published in the *Journal of the American Medical Association* found supplementation with vitamin E reduced the progression of AD.

Research has found a close relationship between vitamin D levels and whether one develops AD. One study found a 122 percent increase in risk to develop dementia for those with low levels of vitamin D. Vitamin D receptors are directly involved with memory, brain nerve growth factor, and protecting the brain from toxins. Vitamin D is also anti inflammatory.

The vitamins folate, vitamin B6 and vitamin B12 can reduce levels of homocysteine, which is linked to tau protein involvement with AD. High levels of homocysteine increase the risk of AD. A deficiency of vitamin B12 can lead to confusion and memory problems. Vitamin B12 slows the accelerated rate of brain shrinkage. A folate deficiency is associated with cognitive decline and increased risk of dementia.

One study suggested by taking 1,000 mg of vitamin C, 20 mg of zinc, and tumeric (curcumin) may help prevent dementia. Vitamin C is an antioxidant and tumeric is anti-inflammatory.

Scientists have discovered that the amino acids L - arginine and L - citruline can increase the blood flow to the brain by simply increasing the levels of nitric oxide, which improves memory.

Herbs

A number of herbs have been found to be beneficial in preventing cognitive decline including the following:

Sage: Sage has been used in the treatment of cerebral vascular disease for over a thousand years. It has been found to provide better brain function; it boosts insulin action, and some believe it may be valuable for treating AD and diabetes.

Ginseng: Ginseng has been found to improve concentration and focus by enhancing mental functions.

Dandelions: Dandelions have been shown in studies to fight cancer, prevent osteoporosis, and treat AD. It contains a substance called alpha-glucosidase, which naturally reduces blood sugar and is used in treating diabetes.

Tumeric (Curcumin): Tumeric has a long history in India as being a healing herb. Studies found curcumin has the ability to cross the blood brain barrier and assist the body to combat dangerous inflammation, the root cause of cognition decline. It encourages the immune system to send macrophages to the brain.

Pomegranate: Pomegranate was found in a 2013 UCLA study of older adults to increase verbal memory performance and to increase functional memory activity in MRI testing. The study involved the subjects drinking just eight ounces of pomegranate juice daily for four years. Pomegranate juice has more antioxidant capacity than red wine, grape juice, and green tea.

CBD Cannabinoids: CBD is a cannabinoid that is thought to be a neuro protector. Researchers showed promise that CBD can limit neurobiological damage from AD.

Marijuana: Marijuana inhibits the enzyme acetylcholinesterase (ACHE) and prevents ACHE induced amyloid beta peptide, thus preventing plaque buildup.

Rosemary Aroma: In a study of 150 people, Rosemary aroma was found to significantly improve both mood and prospective memory.

Cinnamon: In a study at Tel Aviv University, cinnamon was found to inhibit the compound found in plaque formation. The tangle filament found in AD was actually disassembled and eliminated when cinnamon in food was consumed.

Coffee

Hundreds of scientific studies have shown that daily coffee drinking can lower the risk of AD, strokes, heart attacks, obesity, and cancer. One study found that people with mild cognitive impairment were less likely to develop full fledged AD by drinking three cups of coffee daily.

Drinking three to five cups of coffee per day can reduce the risk of developing AD by 20 percent. Coffee is high in polyphenoids, which reduce inflammation and slows the deterioration of brain cells.

Caffeine boosts the effect of the neurotransmitters dopamine and serotonin, as well as enhances the level of acetylcholine, a neurotransmitter that improves short term memory. As little as 4 ounces of drip brewed coffee can boost mood and memory. A 2012 study on 124 subjects with mild cognitive impairment, which often proceeds AD, had memory tests done at the beginning of the study and two to four years later. Those who had caffeine levels consistent with three daily cups of coffee did not develop AD. Those that developed dementia had caffeine levels that were 51 percent lower. An article in the *The Journal of Alzheimer's Disease* concluded that coffee provided the first direct evidence associated with a reduced risk of dementia or delayed onset. Coffee can also reduce the risk of type 2 diabetes, Parkinson's disease, stroke, depression, and heart disease.

Scientists at Stanford University School of Medicine found drinking coffee protects seniors against inflammation associated with AD. Also studies have found a single cup of black, green or oolong tea daily was linked to a 50 percent reduction of developing dementia.

Wine

People who have been diagnosed with AD have an average life span of eight to ten years. A four-year study involving alcohol and AD on 321 subjects found those who drank one to one and half glasses of wine per day had a 77 percent lower risk of death due to AD than those who had a half glass or lower of wine per day.

A study on 550 subjects for more than 30 years found those who drank two glasses of wine or beer each night had the largest mental decline. This amount could cause hippocampus atrophy.

Scientists found that regaining brain function is possible a year after abstaining from alcohol.

Another study found those who did not drink, 35 percent had brain shrinkage on the right side of the hippocampus compared to 77 percent of heavy drinkers, and 65 percent in moderate drinkers.

Research has suggested that red wine may prevent AD from progressing because of the resveratrol in it and it's antioxidants.

People who imbibed more than the equivalent of eight glasses of wine in the United States per week increased their risk of dementia by 40 percent compared to moderate drinkers, while tee totalers had a risk 74 percent higher than moderate drinkers.

Diet and Sugary Soft Drinks

A study published in the *Journal of Alzheimer's and Dementia* found sugary drinks like fruit juice and soda sweetened with sugar were more likely to have poorer memory and smaller brain volume. High sugar consumption has been linked to AD. A study on 4,000 people who consumed up to two sugary drinks daily had reduced brain volume. Those who drank more than two sugary drinks had the equivalent brain function of someone 11 years older.

Another study found adults who drank one or more artificially sweetened drinks daily were almost three times more likely to develop dementia or suffer from a stroke compared to those who did not consume soft drinks. The 17-year study was conducted through Boston University School of Medicine.

Diabetes

People with diabetes double their risk for developing AD. High blood sugar makes AD plaques more toxic to the brain. A Japanese study found that patients with type 2 diabetes or resistance to insulin were more likely to develop brain plaques.

A study published in the *Journal of Alzheimer's and Dementia* found that people in early stage type 2 diabetes have signs of brain dysfunction. They showed a high level of insulin resistance and a reduced ability to use glucose. Chronic high levels of blood sugar increase the risk of diabetes.

A study of 15,000 people with type 2 diabetes who took Metformin, an insulin sensitizer, had a significant reduction in risk of developing dementia compared to a control group. Another study found women without diabetes were twice as likely to have good cognitive function compared to women with diabetes.

Obesity often leads to diabetes. Research has shown that people's weight in middle age may influence not just wether they go on to develop diabetes, but obesity in mid life has long been suspected of increasing the risk of AD. One study found that people who had been obese with a body mass index (BMI) of 30 during middle age, were struck with dementia about a year earlier than an overweight person with a BMI of 28. The threshold of being overweight is a BMI of 25.

Exercise

Exercise helps both the heart and brain. One study found that older adults who practiced regular or intense exercise like running and swimming displayed the memory and cognitive skills of someone a decade younger, compared to those who were sedentary or did light exercise such as gardening. The study involved 900 adults. Exercise also lowers blood pressure, elevated cholesterol, and diabetes.

Research published in the *Journal of Alzheimer's Disease* confirmed that physical exercise in mid-life helps reduce developing cognitive impairment in old age. The Finnish study followed 3,000 twins who were given questionnaires in 1975 and

1981, and a phone interview 25 years later for cognition evaluation. Those who participated in vigorous physical activity in middle age displayed lower levels of cognitive impairment compared to those who did less vigorous exercise.

Exercise produces brain-derived neurotrophic factor (BDNF) vital to brain health

Another study on thousands of older people whose walking pace began to slow and who had cognitive complaints, found they were more likely to develop dementia within 12 years.

Exercise enhances mental ability, stops brain shrinkage, and helps form new neurons. Research has shown that physically active people had lower rates of AD and other associated neurodegenerative disorders.

One study found when previously sedentary men and women, aged 50 to 80, walked around a track 40 minutes a day three times a week, the hippocampus actually increased in size.

Research of 900 men and women with an average age of 71 found those who exercised moderately or rigorously over five years (such as jogging, hiking, swimming, or dancing) performed on par with someone a decade younger on tests of memory and other brain skills The study supported the theory that heart health and brain health were linked together. It helps keep blood vessels healthy, ensuring an optimal flow of blood to the brain. Strength training seems to help by sending pulses of blood into the brain.

Researchers found that squeezing a rubber ball with the left hand and then the right hand activated the frontal lobes. Squeezing the ball increases the memory of words by 15 percent.

A study at UCLA on 876 men and women age 65 or older found those who were most active had a 50 percent reduced risk of developing AD. Exercise is excellent in detoxifying the body through sweating and cleaning out the cells of toxins. If mercury is involved, it might explain why exercise reduces dementia.

Lifestyle

Over consumption of alcohol and smoking have also been associated with dementia and AD. However, a daily glass of red wine has been shown to reduce the risk of AD. Resveratrol in wine has antioxidants that neutralize free radicals associated with AD and mercury toxicity.

Research has shown that 35 percent of all dementia could be prevented. The three most important risk factors were early life education that could prevent dementia later on in life. Reduced hearing in mid-life increases the risk of dementia, and smoking was another factor. Stress is associated with dementia and AD. Stress increases corticotropin releasing factor (CRF), a hormone linked to AD. Stress reducing activities may include walking, yoga, meditation, and playing the piano.

Loneliness may provoke a chronic stress reaction that promotes inflammation, which may contribute to AD. Staying socially active has been shown as a way to keep the brain healthy. Activities such as volunteering, going to church, and developing a network of friends helps reduce the stress of loneliness. A study at John Hopkins Bloomberg School conducted a two-year study which found subjects involved in meaningful social activities did not have brain shrinkage, and the memory centers of the brain grew moderately.

According to a study published in the *Journal of Alzheimer's Disease*, yoga was found to be an effective memory booster and had the effects of making one feel better. The controlled study showed improvement in visual-spatial memory as well as improving symptoms of depression and anxiety. A 2008 study published in the *Journal of Psychological Science* of subjects who spent time outside in a natural park-like environment showed improvement in many cognitive functions, compared to those in a city. Drawing was also found to improve memory. People who ate breakfast had enhanced attention, memory, and creativity compared to those who did not eat breakfast.

Music therapists who work with AD describe patients "waking up" when the sound of "loved and familiar music fill their heads." After months of not talking, they begin to talk more and some begin to remember names long forgotten." Neurologist Oliver Sacks wrote in his book *Musicphilia* that for AD patients, music can be very much like a medicine. Researchers found that people who regularly sang had an increase in cognitive ability.

Staying intellectually active reduces the risk of AD. Activities such as reading books, writing letters, and learning a new language may postpone dementia. Doing crossword puzzles is great for reducing the risk of dementia.

Sauna

A Finnish study published in *Age and Aging* in December 2016 conducted by the University of Eastern Finland found a link between sauna visits and memory disease. The study followed 2,300 middle aged Finnish men for more than 20 years. Men who went to the sauna 4 to 7 times a week were found 66 percent less likely to be diagnosed with dementia, and 65 percent less likely to be diagnosed with AD compared ro men taking a sauna once a week. The men who spent time in the sauna every day were less likely to die of heart problems compared to those who spent one day in the sauna each week.

After a sauna, blood pressure is lowered, and it provides a relaxing effect. Sauna is a good method to detoxify the body of toxins, such as mercury through the sweating process.

Sleep

Studies have shown that people with sleep problems have a higher risk for developing AD. Sleep flushes the brain of toxic synaptic-damaging compounds. Poor sleep may lead to amyloid plaque buildup because of the inability to clear out damaging toxic compounds. Good sleep can delay the onset of AD by five years. A sleep questionnaire given to 101 older people with a genetic risk factor found those with sleep difficulties had more biological markers for AD, such as beta amyloid, tau, and brain cell damage.

Studies have found anyone who regularly sleeps less than six hours a night has a higher risk of depression, psychosis, and stroke.

In AD, sleep disruptions are one of the earliest symptoms. As a result, the biological clock is disrupted, which is linked to AD. Proteins associated with neurological disease diminishes the function of the biological clock. A fragment of the amyloid precursor protein is able to enter the nucleus of cells and interfere with the biological clock. There is an increased production of the enzyme dBace, which reduces the expression of a core biological clock protein. Proper sleep allows neuronal repair activity and maintains neuronal health. The research was done on fruit flies.

Breathing disorders such as sleep apnea puts older people at higher risk for memory problems and dementia. Studies have shown that treating sleep apnea helps delay memory problems and also helps the heart. One study found those who reported sleep apnea began to show signs of mild cognitive impairment at age 77. Those without sleep apnea began to develop memory loss around age 90.

For those who were treated for sleep apnea, the age of onset for memory decline was almost identical to those without sleep apnea.

A study involving 2,500 seniors showed on average that subjects who slept fitfully or diagnosed with sleep apnea developed mild cognitive impairment 10 years earlier than sound sleepers. A lack of sleep causes the accumulation of a sticky protein called beta amyloid protein.

MRI studies have shown that sleeping on your side helps cerebral spinal fluid move freely throughout the brain, meaning it helps clear out more toxins from the brain and reduces the risk of AD.

Mercury toxicity causes sleep disorders including insomnia.

Depression

Up to 40 percent of people with AD also have depression, which may be the first sign of AD. Researchers found those who developed dementia were more likely to experience mood and behavior changes before any other symptom.

In a study involving 2,400 Americans 50 years and older, during the first four years, 30 percent of those who would later go on to develop dementia already showed signs of depression, compared to 15 percent who didn't develop dementia showed depression. Dementia is not a specific disease but a general term used to describe a decline in cognition or mental ability, including memory. Depression has been found to cause changes in the hippocampus.

A study found 23 percent of patients who first developed depression at a mean age of 73, later developed AD. 30 percent of patients who developed depression at a mean age of 78, developed AD.

A Yale School of Public Health study found men and women who viewed aging negatively had a greater loss of hippocampus volume and significantly higher scores of plaque formation in the brain. A 2015 study with college students found that happy moods in general were associated with lower inflammation.

Research conducted on middle age women who were guilt ridden and moody, were twice as likely to develop dementia compared to those who were not.

According to a study conducted over 40 years, women who were introverted were more likely to develop AD. For those people with negative emotions, it would be wise to consult a psychotherapist. One of the many symptoms of mercury toxicity is depression and mental health disorders.

Pollution

Fluoride has been defined as a "poisonous pale yellow gaseous element of the halogen group." It is an enzyme inhibitor that has been linked in several studies to AD, including a 1982 study from the University of Iowa.

The *Lancet Medical Journal* reported on a study that found people who lived 50 meters within high traffic roads had a 7 percent higher chance of developing dementia compared to those who did not live close to a busy road. The study conducted in Ontario analyzed the records of 6.5 million residents aged 20 to 85. They found 243,611 cases of dementia between 2001 and 2012. Researchers mapped proximity to major roadways with postal codes. The risk of dementia went down to 4 percent of people who lived 50 to 100 meters from major traffic and down to 2 percent if they lived within 101 to 200 meters. At more than 200 meters, the elevated risk faded away. Mercury vapor is often found in air pollution.

Blood Pressure

Research found adults who were 55 years and older whose systolic blood pressure (top number) fluctuated by 15 points measured regularly over seven years had a faster cognitive decline. Scientists thought it may be a sign of inflammation.

In 2013, scientists found that older people with high blood pressure were more likely to have biomarkers of AD in their spinal fluid. Several studies have correlated dental amalgam mercury with high blood pressure.

Statin drugs that are used to treat high cholesterol have been associated with memory loss. Beginning treatment with statins increased the risk of developing acute memory loss within 30 days nearly four fold. A similar increase of risk was seen in patients starting non-statin drugs causing acute memory loss.

Gum disease

The *Washington Post* in its Science and Medicine section reported about a study connecting gum disease with AD, heart disease, and diabetes based on a large scale data analysis. In people 65 years and older, nearly 70 percent have some degree of periodontal disease.

Microbes

According to a team of 31 scientists at Oxford and Cambridge universities, they believe AD and dementia are caused by a viral infection. The virus they believe attaches to the brain and remains dormant for years. One should keep in mind that mercury does suppress the immune system against microbes.

Some scientists have linked helicobacteria pylori, a bacteria associated with stomach inflammation to AD. The study found 88 percent of AD patients had the bacteria compared to 47 percent of non-AD subjects.

A 1998 study found 22 of 23 AD brains had chlamydia, compared to 1 in 25 in normal brains.

The herpes simplex virus that causes cold sores, increases the amount of amyloid precursor protein found in AD. One should quickly treat a cold sore with an antiviral agent. Those with herpes have a higher risk of dementia. The herpes virus is known to damage the central nervous system and the limbic region of the brain. There have been around a hundred scientific papers linking herpes simplex and AD.

Several Harvard scientists now believe plaques may form to protect the brain from viruses, bacteria, or other pathogens. Mice studies have confirmed this hypothesis by showing amyloid protects against infection.

Mercury exposure increases herpes replication and increases chlamydia infections. (Refer to chapter 7).

Surgery

The author's mother, who had a normal memory for an 80 year old, went in for back surgery and came out with dementia. Studies have shown that one in three people without a prior history of cognitive problems can acquire dementia during surgery following general anesthesia. The name given to this dementia is hospital acquired delirium. A study of 4,500 men and women found those with the highest levels of vitamin D had the lowest risk of hospital acquired delirium. It is recommended that one take 3000 IU of vitamin D daily for a week or two before surgery.

Women

At age 65, women have a 1 in 6 chance of developing AD compared to 1 in 11 for men. Estrogen therapy after 65 may increase the risk of dementia. Older women with mild impairment worsen about twice as fast as men.

Two thirds of the 5.5 million people 65 years or older with AD are women. Researchers have found the number of children a woman has when entering menopause could affect her chance of developing AD. Women with three or more children had a 12 percent lower risk for dementia than women with one child. Entering menopause at age 45 or younger increases a woman's dementia risk by 28 percent. A study of 15,000 women found a 9 percent greater risk for dementia for each miscarriage a woman endured.

Telomeres

Chromosomes are where all our genetic material is packaged in the form of DNA. Telomeres are very special caps at the end of each chromosome that protect against deterioration.

Telomerase is an enzyme that adds back to DNA the ends that were worn down. Nobel Prize winner Dr. Elizabeth Blackburn reported, "if your genes urge your telomeres to be better maintained, your chance of getting AD is somewhat less." To protect telomeres, one should exercise, do interesting activities, and don't have long- term stress. The more types of exercise the better, such as rigorous exercise, weight lifting, cycling, and walking. Meditation also helps telomeres.

A PROTOCOL TO PREVENT ALZHEIMER'S

Two Alzheimer researchers, Dean Sherzai,M.D., Ph.D. and Ayesha Sherzai, M.D. from Loma Linda University believe 9 out of 10 cases of Alzheimer's disease could be prevented with the right lifestyle. Their protocol was published in the June 2019 issue of *Bottomline Health*. After 15 years of treating thousands of Alzheimer's disease patients and its frequent precursor known as mild cognitive impairment (MCI), they developed a seven step protocol that helps format habits to protect the brain.

Step 1: There are only two supplements worth taking for brain health. They include an omega-3 supplement with at least 250 mg daily of docoabhexaenoic acid (DHA), the most important omega-3 fatty acid for brain health. The other supplement is 500 microgram (mcg) daily of vitamin B12. Also suggested is an algae-derived omega-3 supplement rather than fish oil, which may have mercury or other toxins in it.

Step 2: Next, add a 'superstar' food to you diet, which is a whole food, plant-based diet that reduces brain-damaging inflammation. It will protect the small brain arteries and supply the brain with nutrients and phytochemicals for optimal functioning.

One 'superstar' food group are mushrooms. A *Phytotherapy Research Journal* article reported older people with MCI had improved cognitive function after 16 weeks of taking dried mushroom powder. Take it with your meals two to three times a week. If you are not able to follow a strictly plant-based diet, try a Mediterranean diet or MIND diet, which focuses on leafy greens, nuts, berries, beans, whole grains, fish and poultry.

Step 3: More than 60 percent of the brain is comprised of fat. One needs to consume the right kind of fat, not saturated fat. Plant based fats found in nuts, seeds, avocados and olives, contain omega-3 fatty acids that are critical for brain health. They don't recommend fish because of mercury and other toxic chemicals. If you eat fish, stay with small, low-mercury fish such as sardines and anchovies. Be sure to eliminate processed junk food such as sugary cereal, chips, white bread, processed meats, and canned soups. One should maximize fresh and frozen vegetables, 100 % whole grains, seeds, and nuts.

Step 4: Sleep detoxifies the brain, and microglia brain cells are activated to remove toxins that accumulate during the day. It is recommended one should get seven to eight hours of sleep a night. Avoid foods that disrupt sleep such as sugary foods, high fat food, chocolate, and caffeine containing foods. Stop eating at least three hours before going to sleep and stop drinking fluids at least two hours before bedtime.

Srep 5: One should exercise because it increases klotho, which is a little known anti-aging hormone that protects against cognitive decline. Klotho levels increase after 20 minutes of intense aerobic exercises. The doctors recommend 25 to 30 minutes of intense aerobic exercises four to five days a week such as brisk walking, biking, working out on a treadmill or elliptical.

Step 6: The doctors also recommend "getting rid of clutter and cleaning out the brain." Stress exhausts the brain. A cluttered environment is a source of stress when the office and home becomes disorderly. A clean and orderly space positively impacts cognition.

Step 7: One of the best and enjoyable ways to build cognitive reserve is dancing. When one dances, it activates various parts of the brain such as motor cortex, the parietal lobe, frontal lobe, and occipital lobe. Dance also provides social interaction, which builds cognitive reserve. Learning to play a musical instrument and mastering a new language also build cognitive reserve. Find a dance studio and take classes or buy a DVD and learn to dance at home.

"These seven steps are not complicated and just may prevent dementia."

ALZHEIMER'S DISEASE RATES

The Alzheimer's Association predicts that AD will rise by at least 14 percent in all 50 states. States in the west and southeast are expected to have the largest increase. Alaska is expected to have a 55 percent increase and Iowa the lowest percent increase of 14.1 percent. The Alzheimer's Association estimates that by 2050 the number of Americans with AD could reach 14 million.

However, new studies have shown the prevalence of dementia of those 65 years and older has fallen by 24 percent, but the number of dementia cases is still expected to rise as millions of baby boomers age. The author of that study published in the *Journal of the American Medical Association* partly attributes the decline to improved health and education levels. The author of this book believes this decline is due to the dental profession using fewer dental amalgams. The dementia rates in people older than 65 years fell from 11.6 percent in 2000 to 8.8 percent in 2012, a decline of 24 percent according to a study of 21,000 people. In Europe, there also has been a steady decline in dementia. Fewer people are being exposed to dental amalgam mercury.

SUPPLEMENTS

During the past several years there has been a number of non-prescription supplements that claim to improve memory. Many newspapers carry their full page advertisements. Listed below are a number of supplements and a phone number for some to order. Most of the supplements claim to be based on science but I would recommend going on line and checking out the company. Interestingly, there have been 190 prescription drugs that have been tested and fail to slow the progression of AD.

I Piracetem

In a double blind study of patients with age associated memory impairment, 87 percent of subjects who took Piracetem showed improvement compared to 22 percent of the control group.

II Noopept

Noopept is 1,000 times more powerful than piracetem, which was engineered by the Russians. It is available in the United States without prescription. Noopept increases the spindle like activity of alpha wave function in all tested brain regions. It has potential for treating AD. The supplement has low toxicity choline found in eggs and grapefruit, and the body turns choline into the neurotransmitter acetylcholine. Studies found a combination of both piracetam and choline exhibit memory retention scores several times better than piracetem alone.

III Cognishield

Cognishield contains L-theamine which stimulates the brain to learn and remember and can cross the blood brain barrier. The supplement also contains choline and piracetem.

IV Lipogen PS Plus

This supplement is also called AHP, and contains an ingredient that regroups cells in the part of the brain that stores memories. It improves memory, concentration, and thinking. One study found a 44 percent improvement in memory function. Scientists believe it stimulates cell growth in the hippocampus. Another study found it reversed mental age by 12 years. The active ingredient is phosphatidylserine, which helps lower the stress hormone cortisol. It is the only supplement that has a "qualified health claim for effectiveness for both cognitive dysfunction and dementia by the United States FDA." To order, call 1-800-428-0550.

V Perceptive

Perceptive was developed by Dr. Thomas Shea, the director of the University of Massachusetts Lowell Laboratory for Neuroscience. In addition to improving memory and one's overall mental clarity, studies show Perceptive may reduce cognitive decline to aging.

VI Recalmax

Recalmax has been shown to improve retention of information in 83 percent of subjects compared to 63 percent in a control group. Recall of words was 74 percent for Recalmax users compared to 41 percent for the control. Maintaining thought was 79 percent in the Recalmax group compared to 52 percent in the control. One can order Recalmax by calling 1-800-762-4160.

VI Alpha GPC

This supplement contains glycerlphosphorycholine which is found in the brain. It produces choline found in neurotransmitters. It is safe and has a lasting effect. There have been 13 clinical trials on 4,000 people. Alpha GPC maintains homocysteine levels, reduces inflammation, and builds muscle tissue. To order, call 1-800-852-9958.

VII Neuroplus

Neuroplus is a supplement studied at Stanford University which improves memory, helps grow new brain cells, and reverses mental decline. Over 3,000 published research papers and 64 clinical studies have established this natural compound as boosting learning and improving memory. Studies have shown it to improve memory by 44 percent and boost brain wave activity by 20 percent. There are no harmful side effects. It contains phophatidylserine. To order, call 1-800-966-5743.

As we have shown, there are many factors that can increase the risk of developing AD and dementia. By knowing these risks, one can delay and possibly even prevent dementia. Diet and exercise are two of the primary methods of delaying the onset of AD and dementia. Most of the preventive methods described can neutralize the toxic effects of mercury, and perhaps this is why it may prevent and postpone dementia and AD.

EPILOGUE

So ends my journey into the scientific world that began 40 years ago. I discovered how nutrition can help mental illness and feel that every psychotherapist should be using nutrition to complement their therapy. I also know what caused my loved one to develop a mental illness, as she was unfortunate to have had a mouthful of mercury amalgams. This past year I attended a funeral of a very good friend of fifty years who died of dementia that was not Alzheimer's disease, and she also had a mouth full of amalgams. Besides AD, I am sure many other types of dementia are caused by mercury from amalgams.

This book has described many illnesses that have been associated with amalgam mercury. Controlled mental health studies have shown amalgam mercury to be significantly associated with depression, anxiety, anger, insomnia, suicidal thoughts, poor concentration, poor memory, guilt and irritability. Amalgam subjects smoke significantly more than subjects without amalgams. Controlled studies found that people with schizophrenia and manic depression improve significantly following amalgam removal. Mental health disorders were definitely the most common health risk of dental amalgam mercury. The studies also provided evidence that mercury negatively affected the brain at a young age. Once inside the brain, the toxic poison remains for decades. Research has shown the AD brain is affected 30 to 50 years before a diagnosis can be made.

Besides mental health, we found a host of physical health problems that are caused by amalgam mercury. Ironically, the hair mercury finding with myopia that put me onto the path, showed that amalgam mercury was slowing the progression of myopia, the only positive health effect of amalgam mercury.

Our controlled studies found people with amalgam mercury had 45 percent (p=0.0001) more adverse health symptoms compared to a control without amalgams. Some of the significant symptoms included menstrual problems in females, digestive problems, chest pain, tremors, feeling tired, hay fever, and oral cavity symptoms. When amalgams were removed, 70 percent of their health symptoms improved or were eliminated, with 82 percent of mental health symptoms improving or eliminated.

Our studies found that mercury was associated with higher blood pressure, raising the question whether amalgam mercury is the cause of hypertension as some believe. Subjects with amalgams had significantly lower red blood cell counts, lower hemoglobin, and lower hematocrit. Their thyroid hormone was also lower. Amalgam mercury also appears to affect the immune system as the amalgam subjects had lower T lymphocytes, T-8 suppressor cells, and a correlation of immunoglobulin G with urine mercury.

Our research showed that multiple sclerosis was probably caused by amalgam mercury. Most of the pathological and physiological changes in MS improve after amalgam removal. The subjects with amalgam removed had 35 percent fewer exacerbations of MS symptoms during the previous year than the control with amalgams. ALS and Parkinson's disease have also been associated with amalgam mercury.

This is but a brief summary of amalgam mercury toxicity that we found in our studies. This book provides evidence that amalgam mercury can cause Alzheimer's disease. All the pathological and physiological changes that occur in AD can be explained by mercury toxicity. Evidence was cited that several AD patients recovered following amalgam removal.

The question is what should be done? The dental profession can easily get along without amalgams, as composite fillings have proven a much safer alternative. Several countries have banned dental amalgams, and their AD rate has declined. Many dentists in the United States have curtailed using amalgams, and the AD rate is declining in this country. The dental profession's response to this tragedy has paralleled the tobacco industry by denying there is a problem. Enlightened dentists cannot speak out because they may lose their dental license and thus their livelihood.

State and national legislatures need to enact legislation to ban mercury in dental fillings. Perhaps there needs to be a class action law suit to prohibit the use of dental amalgam. Definitely more research needs to be done to explore other health disorders that amalgam mercury may cause. Too many people have suffered from this tragedy of amalgam mercury toxicity. Hopefully, the economic politics in science and dentistry will not block the prevention of AD and health maladies caused by dental amalgams.

APPENDIX I

Scientific Papers by
Dr. Robert Siblerud on
Mercury Amalgam Toxicity

Health

1. Siblerud R, The Relationship Between Mercury from Dental Amalgam and the Cardiovascular System, The Science of the Total Environment, 99, 1990.

2. Siblerud R, The Relationship Between Mercury from Dental Amalgam and Health, Toxic Substance Journal, 10, 1990.

3. Siblerud R and Kienholz E, Evidence that Mercury from Silver Dental Fillings May Slow the Progression of Myopia, J of Orthomolecular Medicine, 13 (3), 1998.

4. Siblerud, R. and Mutter, J. An Overview of Evidence that Mercury from Dental Fillings may be an Etiological Factor in Many Health Disorders, Journal of Biomedical Research and Environmental Sciences, Jun, 2021.

5. Siblerud R, Health Effects After Dental Amalgam Removal, J of Orthomolecular Medicine, 5 (2), 1998.

6. Siblerud, R, Kienholz, E, and Motl, J. Evidence that Mercury from Silver Dental Fillings may be an Etiological Factor in Smoking, Toxicology Letters, 68, 1993.

7. Siblerud, R, The Relationship between Mercury from Dental Amalgams and Oral Cavity Health, Annals of Dentistry, VXXXIX, N. 2, Winter 1990.

Mental Health

1. Siblerud R, The Relationship Between Mercury from Dental Amalgams and Mental Health, American J of Psychotherapy, XLII (4), 1989.

2. Siblerud R, Motl J, and Kienholz E, Psychometric Evidence that Mercury from Silver Dental Fillings may be an Etiological Factor in Depression, Anger, and Anxiety, Psychological Reports, 74, 1994.

3. Siblerud R, Motl J, and Kienholz E, Psychometric Evidence that Dental Amalgam Mercury may be an Etiological Factor in Manic Depression, J of Orthomolecular Medicine, 13 (1), 1998.

4. Siblerud R, Motl J, and Kienholz E, Psychometric Evidence that Dental Amalgam Mercury may be an Etiological Factor in Schizophrenia, 14 (4), 1999. The Journal of Orthomolecular Medicine, V14, no. 4, 4th Quarter, 1999.

5. Siblerud, R. and Mutter, J, A Hypothesis and Evidence that Mercury May be an Etiological Factor in Schizophrenia, Clinical Schizophrenia and Related Psychosis, V15, Jan 2021.

Multiple Sclerosis

1. Siblerud R and Kienholz E, Evidence that Mercury from Silver Dental Fillings May be an Etiological Factor in Multiple Sclerosis, The Science of the Total Environment, 142, 1994.

2. Siblerud R, A Comparison of Mental Health of Multiple Sclerosis Patients Without Silver/Mercury Dental Fillings and those with Fillings Removed, Psychological Reports, 70, 1992.

3. Siblerud R and Kienholz E, Evidence that Mercury from Dental Amalgams May Cause Hearing Loss in Multiple Sclerosis Patients, J of Orthomolecular Medicine, 12 (4), 1997.

4. Siblerud R and Kienholz E, Evidence that Mercury from Silver Dental Fillings may be an Etiological Factor in Reduced Nerve Conduction Velocity in Multiple Sclerosis Patients, J of Orthomolecular Medicine, 12 (4), 1997.

5. Siblerud R and Kienholz E, A Comparison of Oral Health Between Multiple Sclerosis Subjects with Dental Amalgams and those with Amalgams Removed, J of Orthomolecular Medicine, 14 (2), 1998.

6. Siblerud, R and Mutter, J. A Hypothesis and Additional Evidence that Mercury May be an Etiological Factor in Multiple Sclerosis, J of Multiple Sclerosis, 7 (3), 2020.

Alzheimer's Disease

1. Siblerud, R., Mutter, J., Moore, E., Naumann, J., and Walach, H. A Hypothesis and Evidence that Mercury May be an Etiological Factor in Alzheimer's Disease, International Journal of Environmental Research and Public Health, 16 (24), 2019.

Parkinson's Disease

1. Siblerud, R and Mutter, J. A Hypothesis and Evidence that Mercury May be an Etiological Factor in Parkinson's Disease, Journal of Community Medicine and Health Research, 3(4), 2021.

APPENDIX II

RESULTS

A Comparison of Health and Physiology of Subjects With Dental Amalgams (10 Amalgams Average) to those Without Dental Amalgams Who are Age and Sex Matched

RESULTS

Nonamalgam subjects: 30 females, 21 males
Average age: 22.35 years
Amalgam Subjects: 30 females, 20 males
Average Age: 23.28 years
Average Number of Amalgams: Females 9.8, Males 10.1

Mercury in Hair and Urine (Table 1)
Mercury levels in urine were 201% higher (P=0.0002) in the amalgam group and hair levels were 26.5% higher (P=0.008).

Table 1: Mercury in Tissue

Tissue Mercury	Nonamalgam N=50	Amalgam N=51	Significance (P)
Hair Mercury (ppm)	1.13	1.43	0.008
Urine Mercury (ppb)	1.23	3.70	0.0002

Table 2: Lifestyle Questionnaire (Nonamalgam Subjects, N = 45, Amalgam Subjects N = 47)

A. Mental/Emotional

Mental	Nonamalgam	Amalgam	Significance (P)
1. Stress Tolerance (1 good, 2 avg., 3 poor)	1.76	1.70	0.296
2. Amount of Stress (1 low, 2 avg., 3 high)	2.30	2.41	0.157
3. Emotion Level (1 low, 2 avg., 3 high)	2.13	2.26	0.122
4. Health Description (1 poor, 2 avg., 3 good)	2.54	2.57	0.466
5. Health(scale 1-10, 10 best)	8.36	8.15	0.220
6. Happiness (scale 1-10, 10 best)	8.48	8.02	0.047
7. Peace of Mind (scale 1-10, 10 best)	8.02	7.54	0.033
8. Reading Comprehension Good Average Poor	41 4 0	35 12 0	0.04
9. Grade Point Average (on 4.0 scale)	3.15	3.07	0.196

B. Diet

Diet Category	Nonamalgam	Amalgam	Significance (P)
1. Vitamins	25	28	0.313
2. Minerals	8	14	0.112
3. Fish (2 or more times a week)	4	5	0.487
4. Vegetarian	3	7	0.185
5. Sweet Cravings	13	19	0.14
6. More than 2 cups of coffee daily	9	14	0.175
7. Skip Breakfast	19	20	0.495
8. Eat a lot of sweets	6	10	0.205
9. More than one daily drink of alcohol	3	5	0.375
10. Trouble handling alcohol	1	1	

C. Dental

Dental Category	Nonamalgam	Amalgam	Significance (P)
1.Braces	14	16	0.472
2. Male fillings average:10.1 Female fillings average: 9.8		10.1 9.8	

D. Health

Health Category	Nonamalgam	Amalgam	Significance (P)
1. Yearly Visits to M.D.	1.53	1.79	0.184
2. Birth Control Pill	4	11	0.019
3. Smoking	1	6	0.08

TABLE 3: Health Questionnaire I (Nonamalgam Subjects 45, Amalgam Subjects 47)

A. Mental

Mental Category	Nonamalgam	Amalgam	Significance (P
1. Lack of Interest	3	5	0.375
2. Shyness	14	10	0.20
3. Nightmares	4	7	0.285
4. Forgetfulness	8	10	0.435
5. Lack Self Confidence	5	10	0.15
6. Nervous	8	12	0.26
7. Fear	3	6	0.265
8. Lack of Attention	4	7	0.285
9. Inability to Concentrate	8	12	0.26
10. Loss of Memory	3	3	
Total	63	85	35% More Symptoms in Amalgram Group

B. Dental

Dental Category	Nonamalgam	Amalgam	Significance (P)
1. Foul Breath	0	4	0.07
2. Bleeding Gums	3	8	0.11
3. Grind Teeth	4	7	0.14
Total	7	19	171% More Symptoms in Amalgram Group

C. Health

Health Category	Nonamalgam	Amalgam	Significance (P)
1. Tremors	4	17	0.07
2. Asthma	1	5	0.11
3. Edema	2	2	
4. Loss of Appetite	2	0	0.23
5. Dim Vision	3	3	
6. Allergies	18	22	0.42
7. Headaches	22	31	0.16
8. Headaches a Month	2.77	3.92	0.13
9. Colds and Respiratory Infections (greater than 2 a year)	9	17	0.05
Total	64	101	58% More Symptoms in Amalgam Group

D. Total

Symptoms	Nonamalgam	Amalgam	Significance (P)
Grand Total Questionnaire 1	124	205	0.0001
			56% More Symptoms in Amalgam Group

TABLE 4 Health Questionnaire II (Nonamalgam Subjects 48, Amalgam Subjects 47)

I. Skin Problems

Skin Category	Nonamalgam	Amalgam	Significnce (P)
1. Unexplained Rashes	5	8	0.267
2. Excessive Itching	5	3	0.367
3. Red Flushes of Color	1	4	0.172
4. Rough Skin	3	4	0.488
5. Acne (pimples)	31	23	0.091
6. Other Skin Conditions	4	14	
Total	49	56	0.174
48 Nonamalgam Subjects 47 Amalgam Subjects	1.02 Symptom per Subject	1.19 Symptoms per Subject	14% More Symptoms in Amalgam Group

II. Cardiovascular

Cardiovascular Symptoms	Nonamalgam	Amalgam	Significance (P)
1. Heart or Chest Pain	1	5	0.098
2. Tachycardia (racing heart beat)	0	3	0.127
3. Heart Murmur	5	5	
4. High Blood Pressure	2	1	0.346
5. Low Blood Pressure	4	2	0.346

6. Other Heart Problems	1	0	0.245
Total	13	16	0.204
48 Nonamalgam Subjects 47 Amalgam Subjects	0.27 Symptoms per Subject	0.34 Symptoms per Subject	23% More Symptoms in Amalgam Group

III. Nervous

Nervous Symptoms	Nonamalgam	Amalgam	Significance (P)
1. Multiple Sclerosis	0	0	
2. Bell's Palsey	0	0	
3. Shingles (Herpes Zoster)	0	3	0.116
4. Numbness in Any Part of Body	4	7	0.248
5. Tingling in Any Part of Body	5	5	
6. Epilepsy or Convulsions	0	2	0.232
7. Dr, Said "It's Your Nerves"	4	3	0.50
8. Shakes of hands, feet, head, etc	2	3	0.49
9. Twitching of eyelids or other muscles	15	9	0.131
10. Other Nervous Disorders	1	2	0.353
Total	31	34	0.353
48 Nonamalgam Subjects 47 Amalgam Subjects	0.65 Symptoms per Subject	0.72 Symptoms per Subject	11% More Symptoms in Amalgam Group

IV. Digestion

Digestion Symptoms	Nonamalgam	Amalgam	Significance (P)
1. Diverticulosis	0	1	0.495
2. Ulcers	2	5	0.207
3. Chrohn's Disease	0	0	
4. Grave's Disease	0	0	
5. Indigestion	6	11	0.131
6. Bloated After Eating	6	9	0.271
7. Heart Burn	4	10	0.068
8. Poor Appetite	4	4	
9. Diarrhea	9	13	0.215
Total	31	53	0.041
48 Nonamalgams Subjects 47 Amalgam Subjects	0.65 Symptoms per Subject	1.13 Symptoms per Subject	71% More Symptoms in Amalgam Group

V. Blood Disorders

Blood Symptoms	Nonamalgam	Amalgam	Significance (P)
1, Mononucleosis	5	3	0.368
2. Anemia	3	6	0.231
Total	8	9	0.391
48 Nonamalgam Subjects 47 Amalgam Subjects	0.17 Symptom per Subject	0.19 Symptoms per Subject	12% More Symptom in Amalgam Group

VI. Endocrine System

Endocrine Symptoms	Nonamalgam	Amalgam	Significance (P)
1. Thyroid	0	1	0.495
2. Pancreas	0	0	

	Nonamalgam	Amalgam	Significance (P)
3. Hypothyroidism	1	0	0.495
4. Ovaries	1	1	
5. Testes	1	1	
6. Menstruation - Painful, Too Often, Too Seldom, Etc.	10	16	0.09
7. Hysterectomy	0	0	
8. Tipped Uterus	2	4	0.327
9. Cervical Erosion	0	2	
10. Prostate Problems	0	0	
11. Overweight	7	10	0.279
12. Underweight	1	1	
Total	23	36	0.019
48 Nonamalgam Subjects 47 Amalgam Subjects	0.48 Symptoms per Subject	0.77 Symptoms per Subject	57% More Symptoms in Amalgam Group

VII. Emotional and Mental

Emotional/ Mental Symptoms	Nonamalgam	Amalgam	Significance (P)
1. Sudden Anger	1	10	0.0046
2. Depression	7	18	0.008
3. Wish You Were Dead	2	4	0.327
4. Irritability	9	19	0.018
5. Suicidal Tendencies	1	4	0.173
6. Divorced	0	1	0.495
7. Frequent Anxiety	9	13	0.225
Total	29	69	0.007

48 Nonamalgam subjects 47 Amalgam Subjects	0.60 Symptoms per Subject	1.47 Symptoms per Subject	138% More Symptoms in Amalgam Group

VIII. Annoying Symptoms

Annoying Symptoms	Nonamalgam	Amalgam	Significance (P)
1. Frequent Headaches	5	7	0.364
2. Migraine Headaches	3	3	
3. Trouble Making Decisions	6	10	0.195
4. Dizziness	3	4	0.488
5. Motion Sickness	9	7	0.409
6. Noises in Ears	7	8	0.482
7. Hearing Problems	7	2	0.085
8. Tire Easily	2	11	0.007
9. Tired in Morning	10	25	0.001
11. Insomnia	5	5	0.487
12. Cold Hands/Feet	14	11	0.343
13. Low Body Temperature	6	10	0.192
14. Frequently Feel Cold	8	7	0.495
15. Frequent Night Urination	8	11	0.286
16. Frequent Day Urination	9	10	0.479
17. Chronic Eye Inflammation	0	2	0.233
18. Excessive Sweat	2	2	
19. Swollen Lymph Nodes	2	4	0.237
20. Leg Cramps	4	4	
21. Slow Healing	2	3	0.49
Total	112	149	0.068

48. Nonamalgam Subjects	2.33 Symptoms per Subject		
47. Amalgam Subjects		3.17 Symptoms per Subject	33% More Symptoms in Amalgam Group

IX. Allergies

Allergies	Nonamalgam	Amalgam	Significance (P)
1. Metals	0	2	0.233
2. Food	3	2	0.49
3. Fabrics	1	1	
4. Soaps/ detergents	2	4	0.327
5. Hay Fever	9	19	0.072
6. Other Allergies	14	14	
Total	29	39	0.115
48 Nonamalgam Subjects 47 Amalgam Subjects	0.60 Symptoms per Subject	0.83 Symptoms per Subject	34% More Symptoms in Amalgam Group

X. Diseases

Diseases	Nonamalgam	Amalgam	Significance (P)
1. Arthritis: Rheumatoid, Osteoid	0	2	0.233
2. Bursitis	1	2	0.492
3. Tennis Elbow	1	0	0.495
4. Painful Joints	2	6	0.127
5. Asthma	2	4	0.327
6. Surgery	13	16	0.304
7. Frequent Colds	2	6	0.127
8. Osteomyelitis	0	0	
9. Psoriasis	1	1	

10. Sickle Cell Anemia	0	0	
11. Chronic Anemia	0	0	
12. Kidney Stones	0	0	
Total	22	37	0.021
48 Nonamalgam Subjects	0.46 Symptoms per Subject		
47 Amalgam Subjects		0.79 Symptoms per Subject	68% More Symptoms in Amalgam Group

XI. The Eye

The Eye	Nonamalgam	Amalgam	Significance (P)
1. Wears Glasses for Distance	34	26	0.083
2. Trouble with Night Vision	3	13	0.006
3. Trouble with Night Driving	8	13	0.148
4. Trouble Adjusting to Dark	4	6	0.355
5. Eyes Frequently Bloodshot	7	9	0.375
6. Sensitive to Light	12	14	0.385
7. Eye Strain	3	5	0.344
8. Eyes Tire when Reading	9	8	0.495
9. Total	80	94	0.17
48 Nonamalgam Subjects 47 Amalgam Subjects	1.67 Symptoms per Subject	2.00 Symptoms per Subject	18% More symptoms in Amalgam Group

XII. Dental History

Dental	Nonamalgam	Amalgam	Significance (P)
1. Metallic Taste	0	8	0.0045
2. Burning Sensation in Mouth	0	1	0.495
3. Increased Flow of Saliva	1	1	
4. Peridontal Disease.	0	2	0.233
5. Gum Disease	1	1	
Total	2	13	0.007
48 Nonamalgam Subjects 47 Amalgam Subjects	0.04 per Symptoms Subject	0.28 per Symptoms Subject	555% More Symptoms in Amalgam Group

TABLE 5: Health Questionnaire Summary Totals

Grand Total Symptoms	Nonamalgam	Amalgam	Significance (P)
Questionnaire I	124	205	0.0001
Questionnaire II	429	607	0.005
Total	553	812	0.0001
Symptoms/ Subject Questionnaire I	2.84 Symptoms per Subject	4.21 Symptoms per Subject	
Symptoms/ Subject Questionnaire II	8.94 Symptoms per Subject	12.91 per Symptoms Subject	
Total	11.78 per Symptoms Subject	17.12 Symptoms per Subject	0.0001
			45% Morein Symptoms Amalgam Group

Symptoms Greater in Amalgam Group	78	62.2%	
Symptoms Greater in Nonamalgam Group	19	15.2%	
Symptoms Equal in Amalgam and Nonamalgam Groups	17	13.6%	
No Symptoms in Each Group	11	8.8%	
Total	125 Symptoms		

Physiology

I. Cardiovascular

Studies have shown that mercury has an affinity for the heart and can cause a multitude of electrocardiogram changes. Studies also show that mercury increases blood pressure.

This aspect of the study compared 50 subjects with amalgams (30 females, 20 males, average age 22.3 years) to 51 subjects with no amalgams (30 females, 21 males, average age 22.4 years). The males averaged 10.1 amalgams and females 9.8 amalgams. Blood pressure was tested as were blood samples.

Results

TABLE 6: Significant Difference of Cardiovascular and Blood Parameters

Cardiovascular	Nonamalgam	Amalgam	Significance (P)
Systolic Blood Pressure (50 Amalgams, 51 Nonamalgam Subjects)	100.71 mm/Hg	106.41 mm/Hg	0.0005
Diastolic Blood Pressure (50 Amalgams, 51 Nonamalgams)	58.67 mm/Hg	63.04 mm/ Hg	0.0015
Heart Rate	72.71	69.90	0.074

Hemoglobin (gdl-1)	14.88	14.66	0.016
Hematocrit (% Packed Cell Volume)	43.91	43.15	0.013
Mean Corpuscular Hemoglobulin Conc.	33.98	34.17	0.027
Total Protein (gdl-1)	7.63	7.54	0.098
T Lymphocytes% 40 Females	55.55	53.00	0.045

TABLE 7: Significant Correlations of Mercury with Cardiovascular and Blood Parameters

Blood Parameters	Significant Correlation with Urine Mercury in Amalgam Group (44 Subjects)	Significance (P)
1. Systolic Blood Pressure	- 0.301	0.0233
2. Diastolic Blood Pressure	- 0.304	0.0234
3. Heart Rate	+ 0.299	0.024
4. Hemoglobin	- 0.421	0.003
5. Hematocrit	- 0.393	0.005
6. Red Blood Cell.	- 0.468	0.001
7. Mean Corpuscular Hemoglobulin Volume	- 0.225	0.017 N=44
8.Eosinophils	- 0.219	0.100 N=36

The amalgam subjects reported significantly more heart/chest pains (P = 0.02), tachycardia (P = 0.127), tired in morning (P= 0.001), and tiring easily (P = 0.007).

III. Immunity

The health questionnaire suggested that subjects with amalgams may have a weaker immune system. They reported more colds, hay fever, and respiratory symptoms. Further evidence that the immune system is affected is demonstrated by the stimulation of antibodies, which directly correlates to the number of fillings and urine mercury.

IgG, IgA, and IgE were all significantly correlated to fillings and urine mercury and amalgams. IgG and IgM provide a defense against bacteria, while IgA is found in mucous secretions fighting both bacteria and viruses.

Results

TABLE 8: Immunoglobulin Correlations

Immunoglobulin	Correlation to Amalgams	Significance (P) N=14
IgG	+ 0.614	0.010
IgA	+ 0.713	0.002
IgE	+ 0.503	0.033
	Correlation to Urine Hg	
IgG	+ 0.46	0.049
IgA	+ 0.59	0.014
IgE	+0.70	0.003

The amalgam group had a significantly lower percentage of T lymphocytes (Table 6). Evidence has found that newly placed amalgams can lower T lymphocyte levels. T lymphocytes are an important defense against cancer and viruses.

IV. *The Eye*

The mean refraction of the nonamalgam group was -2.42 diopters of myopia compared to -1.52 diopters in the amalgam group. This 59% difference was significant at the 0.028 level. Visual acuities averaged 20/205 in the nonamalgam group and

20/122 in the amalgam group (P = 0.017). The eye is considered to be part of the brain.

The amalgam subjects had a white visual field of 40.1 degrees compared to 49.1 degrees in the nonamalgam group. The constricted white visual field was significant at the P = 0.013 level. Mercury has an affinity for the rods. Amalgam subjects had significantly more difficulty with night vision which is associated with the rods.

TABLE 9: Visual Field Correlation of Amalgams and Nonamalgam Subjects Combined to Urine Mercury

Visual Field – Combined Amalgam & Nonamalgam Subjects N = 83	Correlation	Significance (P)
White Field	- 0.232	P = 0.007
Green Field	- 0.173	P = 0.059
Red Field	- 0.210	P = 0.029

TABLE 10: Visual Field Correlation of Amalgam Subjects to Urine Mercury

Visual Field - Amalgam Group	Correlation	Significance (P)
White Field	- 0.270	P = 0.038
Red Field	- 0.468	P = 0.001

REFERENCES

Chapter One

1. deLeon, M.J, An Atlas of Alzheimer's Disease, The Parthenon Publishing Group, New York, 1999.

2. Whalley, Lawrence, Understanding Brain Dementia: A Life Course Approach, Columbia University Press, New York, 2015.

3. Turkington, Carol and Galvin James, The Encyclopedia of Alzheimer's Disease, Facts on File, New York, 2003.

4. Glenville, Marilyn, Natural Solutions for Dementia and Alzheimer's: The Ultimate Goal to Prevent, Lifestyle Press, Kent, Tennessee, 2017.

5. Whitehouse, Peter, What You Aren't Being Told About Today's Most Dreaded Disease, St. Martin's Press, New York, 2008.

6. Nordburg, Gunner, Fowler, Bruce, and Lars Friburg, Handbook on the Toxiciology of Metals, 3rd Edition, Elsevier, Oxford, UK, 2007.

7. Null, Gary, Reboot Your Brain, Gary Null Publishing, New York, 2013.

8. Moore, Elaine and Lisa Moore, Encyclopedia of Alzheimer's Disease, McFarland and Company, Jefferson, NC, 2003.

9. Shankle, Rodman and Daniel Amen, Preventing Alzheimer's, The Berkley Publishing Group, New York, 2004.

Chapter Two

1. Nordburg, G., Fowler, B., Nordberg, M., and L. Friburg, Handbook on the Toxicology of Metals, 3rd Edition, Alsevier, Oxford, UK, 2007.

2. Friburg, Lars, Inorganic Mercury, WHO, Geneva, 1991.

3. Suzuki, T, Imura, A., and T. Clarkson, Advances in Mercury Toxicology, Plenum Press, New York, 1990.

4. Erethism, Wikepedia, 2018.

Chapter Three

1. Nakamura, M. and Kawahara, H., Cellular Responses to the Dispersion Amalgams, Journal of Dental Research, 58, 1780- 1790, 1979.

2. Rupp, N.W., Significance to Health of Mercury Used in Dental Practice: A Review, Rep. Council Bureaus. JADA 82:1401-1407, 1971.

3. Hanson, M., Amalgam - Hazards in Your Teeth, J. Orthomolecular Medicine, 12: 194-201, 1988.

4. Ziff, S., Silver Dental Fillings, the Toxic Time Bomb, Aurora Press, New York, 1981.

5. Gay, D.D., Cox, R.O., and Reinhardt, J.W., Chewing Releases Mercury From Fillings, Lancet I, 984-985, 1979.

6. Svare, C.W., Peterson, L.C., Reinhardt, J.W., Boyer, D.B., Frank, C.W., Gay, D.D., and Cox, R.O. The Effect of Removing Dental Amalgam on Mercury Levels in Expired Air, Journal of Dental Research, 60:1668-1671, 1981.

7. Svare, C.W. and Peterson, L.L., et al, The Effect of Removing Dental Amalgam on Mercury Blood Levels, Journal of Dental Research, IADR Abst. #896, 1981.

8. Vimy, M.I. and Lorscheider, F.L., Serial Measurements of Inta-oral Air Mercury: Estimation of Daily Dose from Dental Amalgam, Journal of Dental Research, 64: 1072-1075, 1985.

9. Radics, J., Schwancler, H., and Gasser, F., Die kristallinen komputen der silberamalgam untersuchungen mit der electonischen rontgen mikrosonds, Zahnarztl Welt, 79: 1031, 1970.

10. Pleva, J. Mercury Poisoning from Dental Amalgam, Journal of Orthomolecular Medicine, 12:184-193, 1983.

11. Stortebecker, P., Mercury poisoning from dental amalgams, Stortebecker Foundation for Research, Bioprobe, Orlando, FL, 138, 151-154, 1985.

12. Stock, A., Die chronische quecksilber und amalgam vergistung, Arch Gewerbepath 7, 388, 1936.

13. Hursh, J.B., Clarkson, T.W., Cherian, M.G., Vostal, J.J., and Vander Millie, A., Clearance of mercury (Hg-197, Hg-203), vapor inhaled by human subjects, Archives of Environmental Health, 31:302-309, 1976.

14. Friberg, L. and Vostal, J., Mercury in the Environment, CRC Press, Cleveland, OH, 1972.

15. Nylander, M., Friberg, L., and Birger, L., Mercury concentrations in the human brain and kidneys in relation from dental amalgam fillings, Swedish Dental Journal, 179- 187, 1987.

16. Eggleston, D.W. and Nylander, M., Correlation of dental amalgam with mercury in brain tissue, Res. Ed. 58:704-707, 1987.

17. Patterrson, J., Weissberg, B., and Dennison, J., Mercury in human breath from dental amalgams, Bulletin Environmental Contamination Toxicology, 39:459-468, 1985.

18. Wranglen, G., and Berendson, B.J., Korrosion och metallykydd, Inst.Teknisk, Electrokemi Korrosionslara Kumal Tkniska, Hogskolan, Stockholm, 70, 1983.

19. Mateer, R.S. and Reitz, C.D., Corrosion of amalgam restorations, Journal of Dental Research, 49:399, 1970.

20. Hanson, M., Nothing New Under the Sun, Journal of Orthomolecular Medicine, 12:202-207, 1983.

21. Stock, A., Die gefahrlichkeit des quecksilber-dampfes and der amalgams, Z, Angew. Chemie., 39: 1984.

22. Procedings from conference on amalgam toxicity, Colorado Springs, CO, October 1, 1983.

23. Huggins, H., Mercury, a factor in mental disease?, Journal of Orthomolecular Medicine, 11:3-16, 1982.

Chapter Four

1. Siblerud, R.L., The Relationship Between Mercury From Dental Amalgam and Health, Toxic Substance Journal, 10:425-444, 1990.

2. Siblerud R., Health Effects After Dental Amalgam Removal, Journal of Orthomolecular Medicine, v5, Number 2, 1990.

3. Siblerud, R. and Mutter, J., An Overview of Evidence That Mercury From Dental Fillings May Be An Etiological Factor In Many Health Disorders, Journal of Biomedical Research and Environmental Sciences, June 2021.

Chapter Five

1. Siblerud, R.L. and Kienholz, E., Evidence That Mercury from Silver Dental Fillings May Slow the Progression of Myopia, Journal of Orthomolecular Medicine, v13, Number 3, 1998.

2. Siblerud, R. and Kienholz, E., Evidence that Mercury from Dental Amalgam May Cause Hearing Loss in Multiple Sclerosis Patients, Journal of Orthomolecular Medicine, v 12, Number 4, 1997.

3. Siblerud, R., The Relationship Between Mercury from Dental Amalgam and the Cardiovascular System, The Science of the Total Environment, 99, 1990.

4. Siblerud, R. and Kienholz E., Evidence that Mercury from Silver Dental Fillings may be an Etiological Factor in Multiple Sclerosis, The Science of the Total Environment, 142, 1994.

5. Siblerud, R., and Kienholz, E., Evidence that Mercury from Silver Dental Fillings May be an Etiological Factor in Reduced Nerve Conduction Velocity in Multiple Sclerosis Patients, Journal of Orthomolecular Medicine, V 12, Number 3, 1997.

6. Siblerud, R., A Comparison of Mental Health of Multiple Sclerosis Patients with Silver/Dental Fillings and Those with Fillings Removed, Psychological Reports, 70, 1992.

7. Siblerud, R. And Kienholz, E., A Comparison of Oral Health Between Multiple Sclerosis Subjects with Dental Amalgams and Those with Amalgams Removed, Journal of Orthomolecular Medicine, v 14, Number 4, 1999.

Chapter Six

1. Siblerud, R., Motl, J., and Kienholz, E., Psychometric Evidence that Mercury from Silver Dental Fillings May be an Etiological Factor in Depression, Excessive Anger, and Anxiety, Psychological Reports, 74, 1994.

2. Siblerud, R., The Relationship Between Mercury from Dental Amalgam and Mental Health, American Journal of Psychotherapy, Vol. XLIII, No. 4, October 1989.

3. Siblerud, R., Motl, J., and Kienholz, E., Psychometric Evidence that Dental Amalgam Mercury may be an Etiological Factor in Manic Depression, The Journal of Orthomolecular Medicine, v. 13, Number 1, 1998.

4. Siblerud, R, Mutter, J, Moore, E, Naumann, J, and Walach, H, A Hypothesis and Evidence that Mercury May be an

Etiological Factor in Alzheimer's Disease, International Journal of Environmental Research and Public Health, 16, 2019.

Chapter Seven

[1] Langa, KM (Mar2015) Is the risk of Alzheimer's disease and dementia declining?, Alzheimer's Res Ther, 7 (1) 34.

[2] Breteler MM, Claus JJ, van Duiijin CM, Hofman A (1992) Epidemiology of Alzheimer's, Epidmiol Rev.

[3] Mutter J, Nauman J, Sadagghiani C, Schneider R, Walach H (Oct 2004) Alzheimer's disease: Mercury as a pathogenic factor and apolipoprotein as a moderator, Neuroendocrinology Letters, 25(5).

[4] Bonacker D, Stoiber T, Wang M, Bolt HM, Rhier R, Degan GH, Unger E (2004) Genotoxicity of inorganic mercury salts based on disturbed microtubule function, Arch Toxicol, 78.

[5] Chin-Chan M, Navarro-Yepes J, Quintanilla-Vegas B (Aug 2015) Environmental pollutants as risk factors for degenerative disorders: Alzheimer's and Parkinson's diseases, Front Cell Neurosci, 10:9.

[6] Leong CC, Syed NI, Lorscheider FL (2001) Retrograde degeneration of neurite membrane structural integrity of nerve growth comes following in vitro exposure to mercury, Neuroreport, 12.

[7] Pamphlett R, Kum Jew S (Sep 2016) Age related uptake heavy metals in human spinal interneurons, PloS One, 11(9), e0162260.

[8] Stoiber T, Bonacker D, Bohm K, Thier R, Degan GH, Unger E (2004) Disturbed microtubule function and induction of micronuclei by chelate complexes of mercury(II), Mutation Research, 563.

[9] UNEP (United Nations Environmental Program (Chemicals) (Dec 2002) Global Mercury Assessment UNEP Chemicals Geneva.

[10] Drevnick PE, Lamborg CH, Horgan MJ (Apr 2015) Increase in mercury in Pacific yellowfin tuna, Environ Toxicol Chem, 34 (4). .

[11] Organization WHO (2007) Health risks of heavy metals from long-range transboundary air-pollution, Copenhagen, WHO, Regional Office for Europe.

[12] Geier DA, Geier MR (2006) A meta-analysis epidemiological assessment of neurodevelopment disorder following vaccines administered from 1994 through 2000 in the United States, Neuro Endocrinol Lett.

[13] Moore EA, Moore S (2012) Encyclopedia of Alzheimer's Disease, McFarland and Company, North Carolina.

[14] Monnet-Tshudi F, Zurich MG, Boschat C, Corbaz A, Honegger P, (Apr-Jun 2006) Involvement of environmental mercury and lead in the etiology of neurodegenerative diseases, Rev Environ Health.

[15] Yano K, Hirosawa N, Sakamoto, Y, Katayama H, Moriguchi T, Young KE, Sheen YY, Asaoke K (2002) Aggregations of amyloid beta-proteins in the presence of metal ions, Toxicology Letters, 144.

[16] Zawai NH, Basha MD, Wei W (2002) The influence of lead and mercury on beta-amyloid aggregation and cytotoxicology, Society for neuroscience abstract, Viewer and Itinerary Planner, Abstract No. 688.1.

[17] Kim DK, Park JD, Choi BS (Aug 2014) Mercury induced amyloid-beta (AB) accumulation in the brain is disrupted by disruption of AB transport, J Toxicol Sci.

[18] Alattia JR, Kuraishi T, Dimitrous M, Chang I, Lemaitre B, Fraeing PC (Jul 2001) Mercury as a direct and potent

gamma secretase inhibitor affecting notch processing and development in Drosophila, ASEB J.

[19] Olivieri G, Brack C, Mullen-Spahn F, Stahelin HB, Hermann M, Renard P, Brockhausm M, Hock C (Jan 2000) Mercury induces cell cytotoxicity and oxidative secretion and tau phosphorylation in SHSY5Y neuroblastoma cells, J Neurochem.

[20] Fujimura M, Usuki F, Sawada M, Rostene O, Godefroy D, Takashima A (Nov 2009) Methyl mercury induces new pathological changes with tau hyperphosphorylation mainly through activation of the c-jun-n terminal kinase pathway in the cerebral cortex, but not in the hippocampus of the mouse brain, Neurotoxicology, 30 (6).

[21] Smith PJ, Langdorf GD, Goldberg J (1983) Effects of occupational exposure of elemental mercury on short term memory, Br J Ind Med, 40 (4).

[22] Hrdina PD, Peters DA, Singhai RC (Nov 1976) Effects of chronic exposure to cadmium, lead, and mercury of brain biogenic amines in the rat, Res Commun Chem Pathol Pharmacol.

[23] Oudar P, Cailard L, Fillon G (Oct 1989) In vitro effect of organic and inorganic mercury on the serotonergic system, Pharmacological Toxicology, 65, 245-248.

[24] Basu N, Klenavi K, Gamberg M, Obrien M, Evans D, Scheuhammer AM, Chan HM (Jun 2005) Effect of mercury on neurochemical-binding characteristics in wild mink, Environ Toxicol Chem, 24 (6).

[25] Albrecht J, Matja E (Jun 1996), Glutamate: A potential mediator of inorganic mercury neurotoxicity, Metabolic Brain Disease, 11.

[26] Yanagisawa HI, Nodera M, Wada O, Inducible nitric oxide synthetase expression in mercury chloride-induced acute tubular necrosis, Medline, PMID 9810145.

[27] Yeter B, Deth R, Kuo HC (Sep 2013) Mercury promotes catecholamines which potentiate autoimmunity and vasodilation: Implications for inositol 1,4,5 -triphosphate 23- kinase C susceptibility on Kawasaki syndrome, Korean Circ J, 43 (9).

[28] Rjanna B, Hobson M (Sep 1985) Influence of mercury on uptake of (3H) dopamine and (3H) norepinephrine by rat brain synaptosomes, Toxicol Lett, 27 (1-3).

[29] Kim DK, Park JD, Choi BS (Aug 2014) Mercury induced amyloid-beta (AB) accumulation in the brain is disrupted by disruption of AB transport, J Toxicol Sci.

[30] Dimitrov M, Alattia JR, Lemmin T, Lehal R, Fligier A, Houacine J, Hussain I, Radke F (Aug 2013) Alzheimer's disease mutations in APP but not gamma secretase modulators affect epsilon-cleavage-dependent AICO production, Nature Communications, Published on line doi:10,1038/ ncomms3246.

[31] Park HJ, Jown HS (Mar2013) Mercury induces expression of cyclooxygenase-2 and inducible nitric oxide synthetase, Toxicol Ind Health, 29 (2).

[32] Mann AJ, Auer HE (Jan 1980) Partial inactivation of cytochrome c oxidase by nonpolar mercurial reagents, J Biol Chem, 255 (2).

[33] Shaffi SA (Sep 1995) Sublethal effect of mercury and lead on monoamine oxidase in different regions of the brain in three fresh water teleosts, Rev Esp Fisiol, 51 (3).

[34] Park HJ, Jown HS (Mar 2013), Mercury induces expression of cyclooxygenase-2 and inducible nitric oxide synthetase, Toxicol Ind Health, 29 (2).

[35] Chandradhar D, Raghunathan R, Joshi B (Dec 1980) Effect of mercury compounds on choline acetyltransferase, PubMed.

[36] Falliel-Morel A, Lin L, Sokolowski K, McCandish E, Buckley B, Radke F (Apr2012) N-acetyl cystein treatment reduces mercury-induced neurotoxicity in the developing rat hippocampus, J Neurosci Res, 90 (4).

[37] Warfvinge K, Hansson H, Hultman P (Jun 1995) Systemic auto immunity due to mercury vapor exposure in genetically susceptible vapor exposure and genetically susceptible mice: Dose response studies. Toxicol Appl Pharmocol, June, 132 (2).

[38] Yamamoto M, Khan N, Muniroh M, Motomura E, Yanagisawa R, Matsuyama T, Vogel CF (May 2017) Activation of interleukin-6 and -8 expression by methyl mercury in human U397 macrophages involves reIA and p50, J Appl Toxicol, 37 (5).

[39] El-Fawal HA, Gong Z, Little AR, Evans HL (1996) Exposure to methyl mercury results in serum auto antibodies to neurotypic and gliotypic proteins, Neurotoxicity, 17 (1).

[40] Gardner RM, Nyland JF, Evans SL, Wang SB, Doyle KM, Crainceanu CM, Silbergeld EK (Dec 2009) Mercury induces an unopposed inflammatory response in human peripheral blood mononuclear cells in vitro, Environ Health Perspect, Dec, 117 (12).

[41] Carneiro MF, Morais C, Small DM, Vesey DA, Barbosa F, Gobe GC (Dec 2015) Thimerosal induces apoptotic and fibrotic changes to kidney epithelial cells in vitro, Environ Toxicol,30 (12).

[42] Thomas CJ, Chen Y, Buck DJ (Oct 2011) Chronic inorganic mercury exposure induces sex-specific changes in central TNF and expression. Importance in autism? Neurosci Lett, 504 (1).

[43] Zhao LQ, Shen J, You OC (May2008) Screen of early indicators for renal damage in mercury workers, Sichuan Da Xue Xue Bao Yi Xue Ban, 39 (3).

[44] Kumar SV, Bhattacharya S (2000) In vitro toxicity of mercury, cadmium, and arsenic to platelet aggregation: Influence of adenylate cyclase and phosphodiesterase activity, In Vitro Mol Toxicol, 13 (2).

[45] Frenkel GD, Cain R, Chao SE (Mar 1985) Exposure of DNA to methyl mercury results in the rate of its transcription by mRNA polymerase II, PubMed, DOI 10. 1016/5006-201X (85) 80021 8.

[46] Cordy P, Veiga MM, Al-Saadi S (2011) Mercury contamination from artisanal gold mining in Antioquia, Columbia: The world's highest per capita mercury pollution, Sci Total Environ.

[47] Simonian L (May2017) Alzheimer's disease, 40 years in the wilderness, download citation, 2017, http//seekingalpha. com/article/4071722_Alzheimer's disease40.

[48] Bencko V, Wagner V, Wagner M (1990) Immunological profiles in workers occupationally exposed to inorganic mercury, J Hyg Epidemiol Microbiol Immuno, 34 (1).

[49] Falluel-Morel A, Sokolowski F, Sisti HM,Sisti HM, Shors TJ, Dicicco-Bloom E (2007) Developmental mercury exposure elicits acute hippocampal cell death, reductions in neurogenesis, and severe learning deficits during puberty, J Neurochem, 103 (5).

[50] Haley, BE, Aluminum in vaccines increases thimerosal's activity https://vactruth.com/10/06.

[51] Yin X, Sun JZ, Mei Y (Sep 2008) Effect of Hg 2+ on calcium channels and intracellular free calcium in trigeminal ganglion neurons of rats, Ahonghua Lao Dong, Wei Sheng, 26 (9).

[52] Liugunnar X, Nordberg F, Jin T (Mar 1992) Increased urinary excretion of zinc and copper in mercuric chloride injection in rats, Biometals, 5 (01).

[53] Basun H, Forsell LG, Wetterberg L (Sep1991) Metab and trace elements in plasma Dis Dement Sect, 3 (4).

[54] Constantinidis J (1991) Hypothesis regarding amyloid and zinc in the pathogenesis of Alzheimer's disease potential for preventive intervention, Alzheimer's disease, Assoc Disord..

[55] Olde Rikkert, MG, Verhey FR, Sijben JVV, Bouwan FH, Dautzenberg PL, Lasink M, Sipers WM, VanAssell DZ, VanHees AM, Stevens M, Vellas B, Schellens P (2014) Differences in nutritional status between mild Alzheimer's disease patients and healthy controls, J Alzheimer's Dis, 41 (1).

[56] Alexander J, Thomassen Y, Aaseth J (Jun 1983) Increased urinary excretion of selenium among workers exposed to elemental mercury vapor, J Applied Toxicol, 3 (3).

[57] Kim H, Kim KN, Hwang JY (Mar2013) Relation between serum folate status and blood mercury concentrations in pregnant women, Nutrition, 29 (3).

[58] Siegel BZ, Siegel SM, Correa T,Sekovanic A, Ajvazi M, (Feb 1991) The protection of invertebrates, fish, and vascular plants against mercury poisoning by sulfur and selenium derivatives, Arch Environ Contam Toxicol, 20 (2).

[59] Ahlrot-Westerlund B, Kauppi M (Dec 1995) The presence of mercury and reduced vitamin B12, Heavy Metal Bulletin.

[60] Zeneli L, Sekovanic A, Ajvazi M, Kurti L, Daci N (Feb 2016) Alterations in antioxidants defense system of workers chronically exposed to arsenic, cadmium, and mercury from coal flying ash, Environ Geochem Health, 38 (1).

[61] Agarwal R, Goel SK, Chandra R, Behari JR (Jan 2010) Role of vitamin E in preventing acute mercury toxicity in rats, Environ Toxicol Pharmacol, 29 (1).

[62] Baker-Racine D, Heavy metal poisoning eliminated, http://.www.y2khealthanddetox.com/truthchol http://www.y2khealthand.

[63] Verity, MA, Sarafian T, Pacifici EH, Sevania A (Feb 1994) Phospholipase A2 stimulation by methyl mercury in neuron culture, J Neurochem.

[64] Rotter I, Kosik-Bogacka DI, Dolegowska,Safranow K, Kuczyska M, Laszcznska M (Jun 2016) Analysis of the relationship between the blood concentration of several metals, macro- and micronutrients and endocrine disorders associated with male aging, Environ Geochem Health, 38 (3).

[65] Agrawai S, Flora G, Bhatnager P, Flora SJ (Jun 2014) Comparative oxidative stress, metallothionein induction and organ toxicity following chronic exposure to arsenic, lead, and mercury in rats, Cell Mol Biol, 60 (2).

[66] Halbach S, Ballaton N, Clarkson TW (1988) Mercury vapor uptake and hydrogen peroxide detoxification in human and mouse red blood cells, Toxicol Appl Pharmacol, 96 (3).

[67] Seppanen K, Soiinen P, Salonen Y,Lotjonen S, Laatikainen R ((Nov 2004) Does mercury promote lipid peroxidation?, Biological Trace Element Research, 101 (2), 79.

[68] Olczk K, Kucharz EJ, Glowacki A (Dec 1994) Influence of chronic mercury poisoning upon the connective tissue in rats. I. Effect of mercuric chloride on glycosamine levels in tissue, serum, and urine. Cent Eur J Public Health, 2 (2).

[69] Sakaue M, Mori N, Makita M, Fujishima K, Hara S, Arishna K, Yamamoto M (Jun 2009) Acceleration of methyl mercury - induced cell death of rat cellular neurons by brain-derived neurotrophic factor in vitro, Brain Res, 1273.

[70] Rao MV, Purohit A, Patel T (Apr2010) Melatonin protection on mercury-exerted brain toxicity in the rat, Drug Chem Toxicol, 33 (2).

[71] Cho YM (Aug2017) Fish consumption, mercury exposure, and the risk of cholesterol profiles: Findings from the Korea National Health and Nutrition Examination 2010 -2011, Environ Health Toxicol.

[72] Chaudhury S, Mukhopadhyay B, Bose S, Bhattacharya S (Mar 1996) Involvement of mercury in platelet aggregation-probable mechanism of action, Biomed Environ Sci, 9 (1).

[73] Henriksson J, Tjalve H (May 1998) Uptake of inorganic mercury in the olfactory bulbs via olfactory pathways in rats, Environ Res, 77 (2).

[74] Siblerud RL, Kienholz E, Motl J (1993) Evidence that mercury from silver dental fillings may be an etiological factor in smoking, Toxicology Letters, 68.

[75] Omura Y, Beckman SL (1995) Role of mercury in resistant infections and effective treatment of chlamydia trachomatis and herpes family viral infections, Acupunct Electrother Res, 20 (2 4).

[76] Inman AD, Chin B (2009) Heavy metal exposure reverses genetic resistance to chalmydia-induced arthritis, Arthritis Res Ther, 11 (4).

[77] Black FJ, Bokhutlo T, Somoxa A (Apr2011) The tropical African mercury anomaly: lower than expected mercury concentrations in fish and human hair, Sci Total Environment, 409 (10).

[78] Siblerud RL, Motl J, Kienholz E (1994) Psychometric evidence that mercury from silver dental fillings may be an etiological factor in depression, excessive anger, and anxiety, Psychological Reports, 74.

[79] Choi B, Yeum KJ, Park SJ,Kim KV, Joo NS (Jan 2015) Elevated serum ferritin and mercury concentrations are associated with hypertension, Environ Toxicol, 30 (1).

[80] Bjorklund G, Steiskal V, Urbina MA, Dadar M, Chirumbolo, Mutter J, (2018) Metals and Parkinson's disease: Mechanisms and biochemical processes, Curr Med Chem, 25 (19).

[81] El-Saeed G, Abdel-Maksoud S, Bassyouni H (Jun 2016) Mercury toxicity and DNA damage in patients with Down syndrome, Medical Research Journal, 15 (1).

[82] Rooney JP (2014) The retention time of inorganic mercury in the brain: A systemic review of the evidence, Toxicol and Pharmacol, 274 (3).

[83] Weiner JA, Nylander M (Sep 1993) The relationship between mercury concentrations in human organs and different predictor variables, Sci Total Environment, 138 (1-3).

[84] Harris HH, Pickering IJ, George GN (2003) The chemical form of mercury in fish, Science 301: 1203.

[85] Heintze U, Edwardsson S, Derand F, Birkhed D (!983) Methylation of mercury from dental amalgam and mercury chloride by oral streptococci in vitro, Scand J Dent Res, 91:150.

[86] Fredriksson A, Dencker I, Archer T, Archer T (1996) Prenatal exposure to metallic mercury vapor and methyl mercury provide interactive behavioral changes in rats, Neurotoxicol Teratol, 18.

[87] Pendergrass JC, Haley BE, Vimy MJ, Winfield SA, Lorscheider FL (1997). Mercury vapor inhalation inhibits binding of GTP to tubulin in rat brain similarity to a molecular lesion in Alzheimer's diseased brain, Neurotoxicology, 18 (2).

[88] Hock C, Drasch G, Golombowski S,Muller-Spahn F, Willerhausen-Zonnchen B, Hock U, Growden JH, Nitsch RM (1998) Increased blood mercury levels in patients with Alzheimer's disease, J Neural Transm (Vienna), 105 (1).

Chapter Eight

1. Mutter, J, Curth A, Naumann, J, Deth, R, and Walach, H, Does Inorganic Mercury Play a Role in Alzheimer's Disease? A Systemic Review and an Integrated Molecular Mechanism, J of Alzheimer's Disease, 22, 2010.

2. Mutter, J, Is Dental Amalgam Safe for Humans? The Opinion of the Scientific Committee of the European Commission, J of Occupational Medicine and Toxicology, 6 (2), 2011.

3. Mangelsdorf I, Walach H, and Mutter J, Healing of Amytrophic Lateral Sclerosis: A Case Report, Complementary Medical Research, 24 (3), doi 10.1159/100477397.

4. Saxe SR, Wekstein MW, Kryscio RJ, et al, Alzheimer's Disease, Dental Amalgam, and Mercury, J of the American Dental Association, Feb 130, 1992.

5. Ehmann WD, Markesbery WR, Alauddin M, and Brubaker EH, Brain Trace Elements in Alzheimer's Disease, Neurotoxicty, 7 (1), 1986.

6. Thompson CM, Markesbury WR, Ehmann WD, Regional Brain Trace-Element Studies in Alzheimer's Disease, Neurotoxicology, 9 (1), 1988.

7. Vance DE, Ehmann WD, and Markesbery WR, Trace Element Imbalances in Hair and Nails of Alzheimer's Disease, Neurotoxicity, 9 (2), 1988.

8. Hock C, Drasch G, Golombouski S, et al, Increased Blood Mercury Levels in Patients with Alzheimer's Disease, J Neural Transm, 105 (1), 1998.

9. Cornett CR, Markesbery, and Ehmann WD, Imbalance of Trace Elements Related to Oxidative Damage in Alzheimer's Disease Brain, Neurotoxicology, 19 (3), 1998.

10. Basun H, Forsell LG, and Winblad B, Metals and Trace Elements in Plasma and Cerebrospinal Fluid in Normal Aging

and Alzheimer's Disease, J Neural Transm, Dementia Sect., 3 (4), 1991.

11. Saxe SR, Snowden DA, Wekstein MW, et al, Dental Amalgam and Cognitive Functions in Older Women; Findings from the Nun Study, J of the American Dental Association, 126 (11), 1995.

12. Ely, JAA, Mercury Induced Alzheimer's Disease: Accelerating Incidence? Bull Environ Contamination, Toxicology, 67, 2001. DOI 101007/500128 - 001-0193 9.

13. Haley, Boyd, The Relationship of the Toxic Effects of Mercury to Exacerbations of the Medical Condition Classified as Alzheimer's Disease, Medical Veritas, 4, 2007.

Chapter Nine

1. Huggins, Hal, It's All in Your Head: The Link Between Mercury Amalgams and Illness, Avery Publishing Group Inc., Garden City Park, NY, 1993.

Chapter Ten

1. Warren, Tom, Beating Alzheimer's, Avery Publishing Group Inc., Garden City Park, New York, 1991.

Chapter Eleven

1. Hardy, James, Mercury Free, Gabriel Rose Press, Maitland, FL, 1996.

Chapter Twelve

1. Winblat, Bengt, et al, Dementia Risk: Mid-Life Opportunities to Delay Dementia Onset, Karolinska Institute, Sweden,

2. Fotuhi, Majid, The Memory Loss: How to Protect Your Brain Against Memory Cure and Alzheimer's Disease, McGraw Hill, New York, 2003.

3. Null, Gary, Reboot Your Brain, Gary Null Publishing, New York, 2013.

4. Small, Gary and Vorgan Gigi, The Alzheimer's Prevention Program, Workman Publishing Co., New York, 2011.

5. Shankle, Rodman and Amen, Daniel, Preventing Alzheimer's, The Berkley Publishing Group, New York, 2004.

INDEX

A

Abortion 28, 29
Acetylcholine 89, 94, 106
Acetyl Choline Transferase 89,
    96, 107
Acrodynia 20
African Americans 7, 103, 109
Albertson 35
Alcohol 4, 9, 10
Alfred Stock 36
Allele ix, 14, 15, 90, 118
Allen 88
Allergies 45, 49, 140, 155
Alois 1, 90
Alpha-2 98
Aluminum 90, 98, 108, 125, 231
Alzheimer's disease viii, xii, xiii, 1,
    2, 5, 11, 14, 17, 30, 34, 35, 36,
    53, 69, 85, 87, 89, 90, 91, 92,
    96, 105, 111, 112, 124, 135,
    151, 152, 153, 154, 155, 158,
    167, 171, 188, 195, 196, 226,
    229, 231, 232, 235
Amalgam War 131
American Dental Association 113
    Amyloid 1
Amyloid Beta Protein 14

Amyloid Plaque 16, 114, 182
Amyloid Precursor Protein 93, 106
Amytrophic Lateral Sclerosis 123,
    135
Anecdotal vii
Anger 11, 45, 48, 71, 72, 74, 76, 77,
    78, 83, 84, 85, 195, 234
Anorexia 160
Anti-inflammatory 61, 169, 174
Antioxidants 101
Anxiety 9, 12, 45, 49, 75, 77, 80,
    83, 84, 135, 164, 180, 195, 234
Apathy 3, 4, 18
APOE 4, 6, 7, 11, 15, 97, 114,
    115, 118
Apolipoprotein ix, 7, 14, 15, 226
Arachidonic Acid 90, 101
Arthritis 122, 136, 154, 163, 234
Association xii, 7, 32, 34, 112, 127,
    129, 132, 134, 151, 157, 158,
    173, 190, 236, 237
Autism 28, 29, 85, 120, 121, 230
Autoimmune 26, 66, 139, 143, 155,
    161, 163
Ayesha Sherzai 187

B

Baasch 136
BACE 95, 107
Basilis of Meynert 117 BDNF 102,
    109, 179

Beta Amyloid Protein 7, 8, 9, 12,
    16, 89, 93, 95, 98, 173, 183
Bipolar Disorder 81
Blackburn 187

E

F

G

H

N

O

T

U

V